MF332904

NEUTROPHILS AND ALTERED MATRIX-INDUCED GENERATION OF KALLIKREIN SINGLET OXYGEN

BIOCHEMISTRY RESEARCH TRENDS

BIOCHEMISTRY RESEARCH TRENDS

NEUTROPHILS AND ALTERED MATRIX-INDUCED GENERATION OF KALLIKREIN SINGLET OXYGEN

THOMAS STIEF

nova publishers

New York

NOTICE TO THE READER

Library of Congress Cataloging-in-Publication Data

ISBN: 978-1-63463-460-1

Library of Congress Control Number: 2014955183

Published by Nova Science Publishers, Inc. † New York

CONTENTS

PREFACE

Neutrophils are the cells that protect our lives against fungi, bacteria, or thrombi. In fungal or bacterial sepsis the flowing blood is exposed to unphysiologic matrices that fold factor 12 and prekallikrein into F12a and kallikrein. These two enzymes start the contact system of coagulation (altered matrix (AM) – coagulation). The intrinsic ten-ase, factor 10a, thrombin, and micro-thrombi are formed. The neutrophils answer to increased blood concentrations of kallikrein (also a tissue enzyme) or micro-thrombi with intracellular NADPH-oxidase assembly and secretion of extracellular myeloperoxidase, together these enzymes generate reactive oxygen species (ROS) of the type of singlet oxygen, a selective destructor of "non-self" and light signal emitter. Two very interesting clinical routine assays quantify the pro-oxidative capacity of neutrophils in unsupplemented or supplemented blood (the blood ROS generation assay (BRGA)) and the kallikrein and thrombin-generating capacity of any material (e.g. drinking water, vitamins, drugs, pathogens) that comes into contact with blood (the recalcified coagulation activity assay (RECA)). Especially the RECA is the breakthrough in innovative hemostasis diagnostic.

Therefore, the present book focuses on activation of neutrophils by singlet oxygen or by kallikrein, an important starter enzyme of clinically so relevant altered-matrix coagulation.

These are the book chapters following this preface:

1. Quantification of Micro-Thrombi Induced Blood ROS Generation

Polymorphonuclear neutrophil granulocytes (PMN) are the effector cells of cellular fibrinolysis. If their CD11b/18 receptor recognizes fibrin the neutrophils assemble their NADPH-oxidase and generate large concentrations of H_2O_2 that is converted to singlet oxygen $^1\Delta O_2{}^*$ by secreted myeloperoxidase. $^1\Delta O_2{}^*$ is strongly anti-thrombotic. Fresh micro-clots enhanced 2 µg/ml ZyA - stimulated ROS generation at 15 min reaction time in fresh citrated blood up to 550fold, in fresh EDTA-blood up to 6000fold. The greatest increase of ROS generation in citrated blood appeared at 25-50% micro-clots, in EDTA-blood at 25% micro-clots. For optimal stimulation the CD11b/18 receptor of the neutrophils binds preferably to diluted clots where the fibrin D - subunit (in EDTA presence without Ca^{2+}) is not hidden by unspecific proteins. Micro-thrombi at 0 µg/ml ZyA-stimulated blood did not enhance blood ROS generation. Thus, fresh micro-clots are not triggers they are primers of blood ROS generation.

2. Systemic Intravascular Coagulation

The rapid laboratory diagnosis of the pre-phase of pathologic (disseminated) intravascular coagulation (PIC phase 0) is a great clinical challenge. Typical PIC-1 means multiple micro-thrombi that obstruct vital organ areas, and that can cause multi-organ failure. There is clinical need for reliable hemostasis parameters to diagnose PIC-0 or PIC-1 to start the adequate treatment in time.

The routine hemostasis tests such as prothrombin time and activated partial thromboplastin time are too blunted; furthermore, thrombocytes might be dysfunctional, cell fragments might be counted as platelets, platelets-count or fibrinogen concentration are acute phase parameters that might increase instead of decrease, and antithrombin-3 activity might change only slightly or it might be altered in an artificially diluted test matrix. In the present review new routine parameters for diagnosis of the early phases of PIC are presented: basal chromogenic thrombin activity (systemic F2a), thrombin generation after plasma recalcification (RECA), F2a generation after intrinsic activation of plasma (INCA), or F2a generation after extrinsic activation of plasma (EXCA), fibrinogen function+antigen+ratio, undiluted antithrombin-3 activity, and active endotoxin (here clinically understood as active lipopolysaccharide or active ß-glucan). A combination of these new assays allows to diagnose the pre-phase of PIC within minutes. Especially the functional ultra-specific systemic F2a - Test (normal range 100±20%; 100%=5.5 mIU/ml) or the therefrom derived RECA helps to distinguish the normal systemic intravascular coagulation (NIC) from PIC-0 (121-150% of normal basal F2a) or from the PIC-1 (151-200% of normal basal F2a). All these simple photometric assays do not require special hemostasis machines to test the dynamic coagulation state and to differenciate between LPS- and ß-glucan- sepsis of each individual patient especially during the "*first six golden hours*" of the disease.

3. Ultraviolet a Photons in the Late Phase of Oxidative Burst

Blood neutrophils communicate by photons. They generate and perceive light of wavelengths in the range of about 300-400 nm (ultraviolet to blue). Is there a temporal relationship between oxidative burst and appearance of certain photons ? The longer the oxidative burst lasts the more photons are generated that do not pass to the neighbor well (NW). In unstimulated blood NW about half the photons are found compared to empty NW, the erythrocytes seem to absorb these photons. In strongly illuminated blood samples the photon receptors are damaged, subsequent triggering results in decreased blood ROS generation. Slightly illuminated blood samples are 2-4fold stimulated (primed). Physiological light upregulates light!

4. Kallikrein Modulates Singlet Oxygen Generation

Kallikrein is a very important starter enzyme of intrinsic hemostasis. Neutrophils react towards altered matrices with activation of their NADPH-oxidase and generation of reactive oxygen species (ROS). H_2O_2 is the original ROS that is converted via myeloperoxidase to the non-radicalic singlet oxygen ($^1\Delta O_2^*$). The regulation of $^1\Delta O_2^*$ generation by kallikrein is the purpose of this work. The approx. IC50 was 0.4 ng/ml kallikrein at BRGA-26 (blood ROS generation assay with 26 min reaction time) for fresh citrated and EDTA-blood. At BRGA-146 appeared an approx. SC200 of 3 ng/ml kallikrein for fresh citrated blood. Thus, at short incubations kallikrein is an inhibitor and at long incubations kallikrein acts as stimulator of the neutrophil´s singlet oxygen generation. This reaction is similar to that of the contact phase activator glucose. First the up-regulation of NADPH-oxidase assembly is hindered by kallikrein, possibly by cell hormone action of kallikrein/C1-inactivator complexes, later the down-regulation of the assembled NADPH-oxidase is inhibited.

5. Metamizole Is a Strong Trigger of Intrinsic Coagulation

Metamizole is an important analgetic and antipyretic drug. It contains a sulfonate group and a benzene ring, i.e. the molecule has strongly negatively charged and strongly lipophilic parts. Negatively charged or lipophilic molecules trigger intrinsic hemostasis, folding factor 12 or pre-kallikrein into F12a or kallikrein. The best assay to determine even spurious activations of contact phase coagulation is the recalcified coagulation activity assay (RECA). Metamizole triggered intrinsic coagulation with approx. SC200 values of 4±2 mg/l. This is a very small concentration, considering that plasma concentrations $>$ 100 mg/l are therapeutically possible. This means that metamizole must be considered a pro-thrombotic risk factor in patients with further thrombosis risks. Then the individual patient who needs metamizole could benefit from LMWH-prophylaxis.

6. Vitamin B₁ (Thiamine) Triggers AM-Coagulation

Thiamine is the hydrophilic vitamin B_1 that consists of amino-pyrimidine linked to thiazole. The normal blood concentration of vitamin B_1 is 20-100 µg/l, most of the vitamin is located inside the erythrocyte. Vitamin B_1 can clinically be given as 100 mg injections. Therefore, this work quantified the changed thrombin generation by alteration of matrix through an increase of vitamin B_1 concentration. Thiamine triggered AM-coagulation with approx. SC200 values of 2.0±1.6 mg/l. This is about tenfold less than the blood plasma concentrations reached directly after injection of 100 mg thiamine. This means that vitamin B_1 should be considered a pro-thrombotic risk factor in patients with further thrombosis risks. Then the individual patient who needs thiamine could benefit from oral administration or a combined vitamin B_1/LMWH treatment.

7. Vitamin B₂ (Riboflavin) Triggers AM-Coagulation

Riboflavine (vitamin B_2) is the central component of the cofactors flavin adenine dinucleotide (FAD) and flavin mononucleotide (FMN). Since riboflavine contains 2 delta-negatively charged oxygens in keto-position, 4 delta-negatively charged oxygens in hydroxyl-groups, and a lipophilic benzene ring with 2 methyl-groups, riboflavin could trigger altered matrix (AM) - coagulation; negatively charged or lipophilic molecules fold factor 12 into

F12a or pre-kallikrein into kallikrein. The best assay to determine even spurious activations of AM-coagulation is the recalcified coagulation activity assay (RECA). The approx. SC200 on AM-coagulation was 0.1±0.04 mg/l (MV±1SD) vitamin B2 (range: 0.04-1.5 mg/l) in 5 of 6 samples. The 6th sample had an approx. IC50 of 0.04 mg/l instead of an approx. SC200. An injection of 10 mg vitamin B2 would instantaneously result in a plasma concentration of about 3 mg/l vitamin B2. This is 30fold the mean value of the approx. SC200 on AM-coagulation in 5 of 6 samples. Therefore, vitamin B2 should not be administered intravenously, oral vitamin B2 has a much slower increase of plasma concentration. If there are indeed clinical situations where vitamin B2 injection is necessary, thiamine should be accompanied by low-molecular-weight-heparin (LMWH) thrombosis prophylaxis.

8. Vitamin B_6 Triggers AM-Coagulation

Vitamin B_6 in its molecular forms pyridoxine, pyridoxal, pyridoxamine and in its active form pyridoxal phosphate is an important cofactor in many reactions of amino acid metabolism. Pyridoxine with its 3 delta-negatively charged oxygens could trigger intrinsic coagulation. The approx. SC200 on AM-coagulation was 1.6±1.0 mg/l (MV±1SD) pyridoxine in 4 of 6 samples. 2 of 6 smples had an approx. IC50 of 0.7 mg/l, in these 2 plasmas vitamin B6 seems to stop AM-coagulation by blocking protein/protein interactions acting like an AM-coagulation interrupter. Vitamin B_6 should not be administered intravenously, oral vitamin B6 has a much slower increase of plasma concentration.

9. Vitamin B_{12} Triggers AM-Coagulation

Cobalamin is the hydrophilic vitamin B_{12} of a complicated structure with a central cobalt ion for normal erythropoiesis and nerve function. Vitamin B_{12} can clinically be given as 1-1.5 mg injections. Therefore, this work quantified the changed thrombin generation by alteration of matrix thru an increase of vitamin B_{12} concentration. The approx. SC200 in AM-coagulation was 19±14 µg/l (MV±1SD) hydroxocobalamin acetate in 5 of 6 plasmas; one plasma had an approx. IC50 of 8 µg/l. Cyanocobalamin had an approx. SC200 of 101±130 µg/l in 3 of 6 plasmas and an approx. IC50 of 13±14 µg/l in 3 of 6 samples. This means that there are intrinsic coagulation differences between hydroxocobalamin acetate and cyanocobalamin. Hydroxocobalamin acetate triggers the contact phase of human coagulation much more than cyanocobalamin. This might be due to the acetate part of the hydroxocobalamin drug; of pharmacologic interest could be to substitute acetate in hydroxocobalamin by HCl.

10. Dramatic Increase of Blood ROS Generation by Vitamins B_1, B_2, B_6, B_{12}

The vitamins B_1, B_2, B_6, B_{12} act as coenzymes of important enzymes in neutrophil physiology. Up-regulation of the NADPH-oxidase, the physiologic generator system of large quantities of reactive oxygen species (ROS; H_2O_2 being the primary ROS and $^1\Delta O_2^*$ being the most selective ROS against "non-self") is central for neutrophils upon contact with fungi, bacteria, or micro-thrombi. The regulation of $^1\Delta O_2^*$ generation by the vitamins B_1, B_2, B_6, B_{12} is the purpose of this work. In the BRGA: BRGA-60- the mean approx. SC200 values of citrated blood were 12:3 mg/l vitamin B_1, 0.6:0.17 mg/l vitamin B_2, 4:5 mg/l vitamin B_6, 0.1:0.1 mg/l vitamin B_{12}. The BRGA:BRGA-60- approx. SC200 values of EDTA blood were 2:10 mg/l vitamin B_1, 0.1:0.5 mg/l vitamin B_2, 2:4 mg/l vitamin B_6, 0.05:0.1 mg/l vitamin B_{12}. Seldom, there occurred an approx. IC50 instead of an approx. SC200. The vitamins studied react differently from individual to individual. Only by performing individual BRGA or BRGA-60- analysis the individual patient will know his personal vitamin dosage that stimulates or suppresses his neutrophils. This information is very important because without knowing the different individual reaction types a patient prone to hyper-inflammation could erroneously take a neutrophils-stimulating vitamin or a patient susceptible to hypo-inflammation could ingest a neutrophils-suppressor.

11. Drinking Water or Beverage Analysis for Intrinsic Coagulation Activation

Alteration of blood matrix is an important trigger of the intrinsic hemostasis cascade. The ingestion of large quantities of beverages changes blood matrix. Therefore, a routine test was established to analyze the action of plasma dilution by different drinks. 0.9% NaCl had the best action on thrombin generation. Addition of 5 µl volume to 40 µl plasma did not change thrombin generation. Drinking water 1 was much better than drinking water 2; at 2 µl supplementation DW1 had 118% thrombin generation, whereas DW2 had 188% thrombin generation. This thrombin activity was comparable to that induced by Pepsi Cola light (assuming that the beverage is intestinally totally absorbed). In conclusion, the present easy technique allows to judge the quality of drinking water or beverages with respect to intrinsic thrombin generation.

12. The True PAP Concentration in Plasma

Plasmin-antiplasmin (PAP) is a biomarker for systemic fibrinolysis activation. The reliable quantification of PAP by enzyme immuno assay (EIA) diagnoses the state of systemic fibrinolysis activation. PAP tests are troublesome because fibrinolysis is extremely difficult to stabilize completely. Whereas coagulation is routinely stabilized by withdrawal of calcium ions and arginine concentrations > 200 mM, fibrinolysis is activated by Ca^{2+} withdrawal and arginine is needed at final concentrations > 600 mM. Here PAP is determined for the first time under completely stabilizing conditions. The PAP concentrations were (hematocrit corrected) 1.6 ng/ml in 728 mM arginine EDTA plasma, 4.8 ng/ml in 315 mM arginine EDTA plasma, 11.2 ng/ml in 0 mM arginine EDTA plasma. The second control plasma with 1.25 M arginine stabilization 1h after blood withdrawal had 4.9 ng/ml PAP. To obtain the true *in vivo* basal PAP conc. EDTA blood must immediately be stabilized by at least about 700 mM arginine. The true 100% normal basal PAP concentration seems to be very close to the true 100% normal basal TAT concentration that is about 1.3 ng/ml.

13. Stimulation of Blood Singlet Oxygen Generation by Serine Proteases

Proteases are very important regulators of cell function. The up-regulation of the NADPH-oxidase, the physiologic generator system of large quantities of reactive oxygen species (ROS; H_2O_2 being the primary ROS and $^1\Delta O_2^*$ being the most selective ROS against "non-self") is the central activity of neutrophils upon contact with fungi, bacteria, or microthrombi. The regulation of $^1\Delta O_2^*$ generation by serine proteases is the purpose of this work. Urokinase stimulated blood ROS generation about 5fold stronger than t-PA in citrated blood. The approx. SC200 were 17 IU/ml u-PA, approx. SC150 = 10 IU/ml u-PA or 300 IU/ml t-PA (6.3 IU t-PA are equivalent to 1 IU u-PA). The approx. SC200 for plasmin in citrated blood was 2 mIU/ml. The approx. SC200 was 2.5 mIU/ml F9a, 10 ng/ml trypsin, or 0.2 ng/ml thrombin. Kallikrein inhibited the singlet oxygen generation with an approx. IC50 of 3 ng/ml and F12a had an approx. IC50 of 19 ng/ml.

In EDTA-blood the 200% stimulatory concentrations were 30 IU/ml u-PA, 300 IU/ml t-PA, 2.5 mIU/ml plasmin, 0.3 mIU/ml F9a, 12 mUml F10a, 10 ng/ml F11a, 10 ng/ml trypsin, 0.6 mIU/ml thrombin. 35 ng/m F12a, 4 ng/ml kallikrein. The approx. SC200 values were in the BRGA-60- version of the assay (citrated blood) were 4 IU/ml urokinase, 2.5 mIU/ml plasmin, 15 ng/ml trypsin, 3 ng/ml kallikrein, 6 ng/ml F12a, 6 mIU/ml F9a, 3 mU/ml F10a, or 0.5 mIU/ml F2a. All serine proteases seem to upregulate the NADPH-oxidase assembly. Only kallikrein and F12a, the starter enzymes of intrinsic coagulation, in the initial phase of neutrophil activation behaved as inhibitors of NADPH-oxidase assembly.

14. Applied Biochemistry of the BRGA

The blood ROS generation assay (BRGA) quantifies the production of reactive oxygen species (ROS) by blood neutrophils. The BRGA screens patients for hypo- or hyper-functional neutrophils, that are dangerous for the individual because of fungal/bacterial sepsis or autoaggression, respectively. There are up to now no informations about the BRGA under routine conditions.

There was no correlation between t-max (73±21 min) and the maximal ROS generation (5158±3924 RLU/s), r= -0.152. The blood ROS generation in BRGA-20 (about 20 min reaction time) was 78±122 RLU/s, in BRGA-40 (about 40 min reaction time) it was 1104±1380 RLU/s, r=0.568. Of particular interest is the ROS generation at about 50% of the time to reach the maximum (0.5t-max; here BRGA-40). BRGA-40 correlates with r=-0.430 with t-max. At 0.5t-max about one third of the samples were pathological, half of them hypo- and half of them hyper-active. It is suggested to screen all patients for the function of neutrophil NADPH-oxidase.

15. Neutrophils in Fibrinolysis

The antithrombotic action of activated polymorphonuclear neutrophil granulocytes (PMN) are reviewed in this book preface. PMN with about 4000 cells/µl are the main leukocyte type in normal human blood. They are the first line of defence against invading fungi or bacteria. About 50 % of all blood PMN are slightly activated and express adhesion molecules (e.g. CD11b/CD18 – Mac 1) that bind endothelial surface proteins (e.g. ICAM-1). PMN roll over the endothelium to patrol for the existence of any pathogen that may stick to the endothelium and that may threaten vessel patency. In this sense also fibrin might be considered as a special type of pathogen. CD11b/CD18 receptors of PMN bind to fibrin and activate the cell. Activated PMN via membrane assembly of NADPH-oxidase (Nox2) produce high concentrations of H_2O_2 that generates HOCl via released myeloperoxidase. Hypochlorite reacts with amino groups (e.g. of taurine) to chloramines, an important biological source of the excited nonradical excited oxidant singlet molecular oxygen ($^1O_2{}^*$). $^1O_2{}^*$ is an inter- and intra-cellular communication signal and the selectively destructive "bullet". Light quants (hv, especially those at about 300-400 nm) enhance PMN patrolling and direct and recruit PMN to a pathologic focus, such as a pathologic thrombus. CD11b/CD18 binding induces H_2O_2 generation and viceversa.

$^1O_2{}^*$ is a selective oxidant. Due to its nonradical nature, $^1O_2{}^*$ is relatively inoffensive to normal mammalian cells but highly destructive against pathogens. $^1O_2{}^*$ inactivates important components of hemostasis, such as thrombocytes, fibrinogen, factors 5, 8, 10, α2-antiplasmin, and PAI-1 with 50 % inhibitory concentrations of 1-2 mM in blood or plasma. It is suggested to imitate the physiologic PMN mediated thrombolysis by intravascular injection of violet photons or of sub-millimolar concentrations of chloramines.

Thomas Stief, M.D., Priv.-Doz.

Institute of Laboratory Medicine and Pathobiochemistry
University Hospital
D-35043 Marburg
Germany
Email: thstief@t-online.de
Tel. +49-6421-58-64471

QUANTIFICATION OF MICRO-THROMBI INDUCED BLOOD ROS GENERATION

ABSTRACT

Background: Polymorphonuclear neutrophil granulocytes (PMN) are the effector cells of cellular fibrinolysis. If their CD11b/18 receptor recognizes fibrin the neutrophils assemble their NADPH-oxidase and generate large concentrations of H_2O_2 that is converted to singlet oxygen $^1\Delta O_2{}^*$ by secreted myeloperoxidase. $^1\Delta O_2{}^*$ is strongly anti-thrombotic.

Material and Methods: Freshest individual normal citrated plasma was recalcified. Immediately thereafter 100 µl amounts (100%, or with 0.9% NaCl diluted to 50 %, 25%, or 12.5%; 0% = no clot) were pipetted into a black polystyrene microtiter plate (Brand®781608). After 60 min (37°C) 120 µl Hanks´ Balanced Salt Solution (HBSS; modified without phenol red), 10 µl freshest normal citrated blood, or 10 µl freshest 1.6 mg/ml K_3-EDTA blood of the same donor, 0.3 mM luminol and 0 or 2.4 µg/ml zymosan A (ZyA) were added. The luminescence was measured by a photons-multiplying microtiter plate luminometer with an integration time of 0.5s.

Results and Discussion: Fresh micro-clots enhanced 2 µg/ml ZyA - stimulated ROS generation at 15 min reaction time in fresh citrated blood up to 550fold, in fresh EDTA-blood up to 6000fold. The greatest increase of ROS generation in citrated blood appeared at 25-50% micro-clots, in EDTA-blood at 25% micro-clots. For optimal stimulation the CD11b/18 receptor of the neutrophils binds preferably to diluted clots where the fibrin D - subunit (in EDTA presence without Ca^{2+}) is not hidden by unspecific proteins. Micro-thrombi at 0 µg/ml ZyA-stimulated blood did not enhance blood ROS generation. Thus, fresh micro-clots are not triggers they are primers of ROS generation.

Keywords: Micro-thrombi, reactive oxygen species, ROS, singlet oxygen, NADPH-oxidase, CD11b/18, neutrophils

INTRODUCTION

Fibrinolysis has a plasmatic part and a cellular one [1-3]. The neutrophils (PMN) are the main cells of cellular fibrinolysis [4]. Their CD11b/18 receptor recognizes fibrin [5]. Then the neutrophils assemble their NADPH-oxidase and generate large concentrations of H_2O_2 that is converted to singlet oxygen $^1\Delta O_2^*$ by secreted myeloperoxidase [6]. $^1\Delta O_2^*$ is strongly anti-thrombotic [7-11].

MATERIAL AND METHODS

Freshest individual normal citrated platelet poor plasma (less than 0.5h old; venous blood of a healthy donor after written informed consent drawn and supplemented with 11 mM sodium citrate in polypropylene monovettes from Sarstedt, Nümbrecht, Germany) was recalcified: 10 parts plasma were mixed with 1 part 250 mM $CaCl_2$ (Sigma, Deisenhofen, Germany).

Immediately thereafter 100 µl amounts (100%, or with 0.9% NaCl diluted to 50 %, 25%, or 12.5%; 0% = no clot) were pipetted into a black polystyrene microtiter plate (Brand, Wertheim, Germany; article nr. 781608). After 60 min (37°C) 120 µl Hanks′ Balanced Salt Solution (HBSS; modified without phenol red; Sigma), 10 µl freshest normal citrated blood, or 10 µl freshest 1.6 mg/ml K_3-EDTA blood of the same donor, 0.3 mM luminol sodium salt in 0.9% NaCl (Sigma) and 0 or 2.4 µg/ml zymosan A (ZyA) (Sigma) were added. The luminescence was measured by a photons-multiplying microtiter plate luminometer (LUmo; Autobio-anthos, Krefeld, Germany) with an integration time of 0.5s per well [12,13].

RESULTS AND DISCUSSION

Fresh micro-clots enhanced 2.4 µg/ml ZyA - stimulated ROS generation at 15 min reaction time in fresh citrated blood up to 550fold (Figures 1,2), in fresh EDTA-blood up to 6000fold (Figures 3,4). The greatest increase of ROS generation in citrated blood appeared at 25-50% micro-clots, in EDTA-blood at 25% micro-clots. The second increase of blood ROS generation in citrated blood is due to activation of cellular fibrinolysis, the recalcified blood generates an additional micro-clot [14].

100% micro-clots often inhibited blood ROS generation, for optimal stimulation of blood ROS generation the CD11b/18 receptor of the neutrophils binds preferably to diluted clots where the fibrin D - subunit (in presence of a S100-protein or EDTA without Ca^{2+}) [15-23] is not hidden by unspecific proteins (e.g. albumin) [24,25].

Few specific triggers and many primers stimulate blood ROS generation. Micro-thrombi at 0 µg/ml ZyA-stimulated blood did not enhance blood ROS generation. Thus, fresh micro-clots are not triggers, they are primers of blood ROS generation.

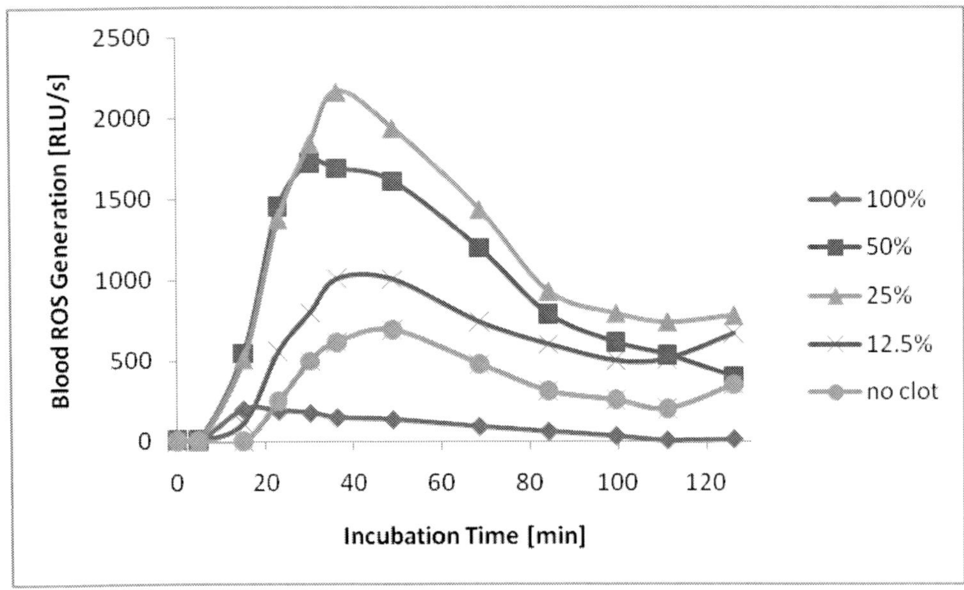

Figure 1. ROS generation induced by fresh plasmatic micro-thrombi in fresh citrated blood. Fresh normal citrated platelet poor plasma was recalcified and diluted to 50%, 25%, 12.5%. Immediately thereafter 100 µl portions were pipetted into a black polystyrene microtiter plate (Brand®781608). After 60 min (37°C) the 2.4 µg/ml zymosan A stimulated ROS generation in citrated blood was determined as described in Methods. The second increase of ROS generation in no clot controls and 12.5% clots is due to blood micro-thrombi by recalcification during the BRGA [14].

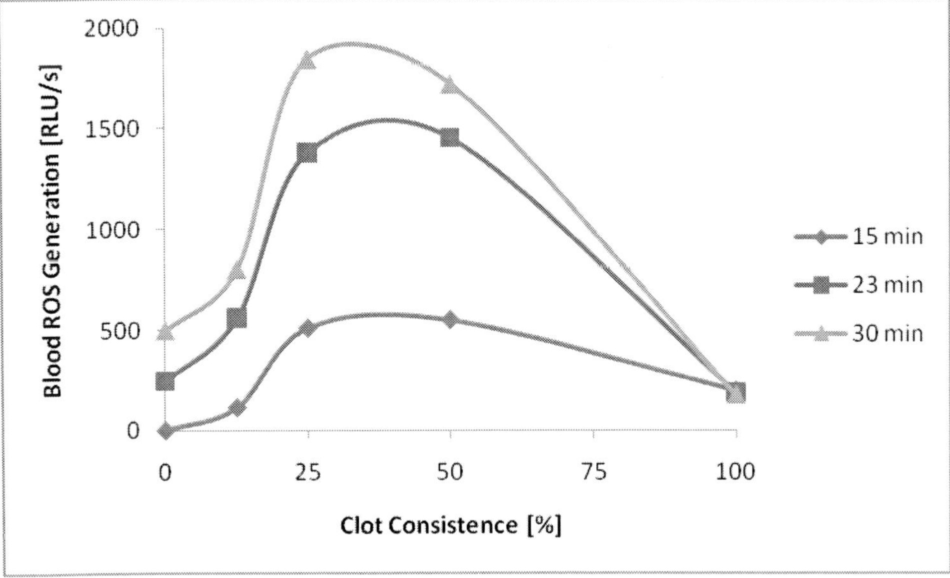

Figure 2. ROS generation depends on clot consistence. Fresh normal citrated platelet poor plasma was recalcified and diluted to 50%, 25%, 12.5%. Immediately thereafter 100 µl portions were pipetted into a black polystyrene microtiter plate (Brand®781608). After 60 min (37°C) the 2.4 µg/ml zymosan A stimulated ROS generation in citrated blood at 15, 23, 30 min was determined as described in Methods.

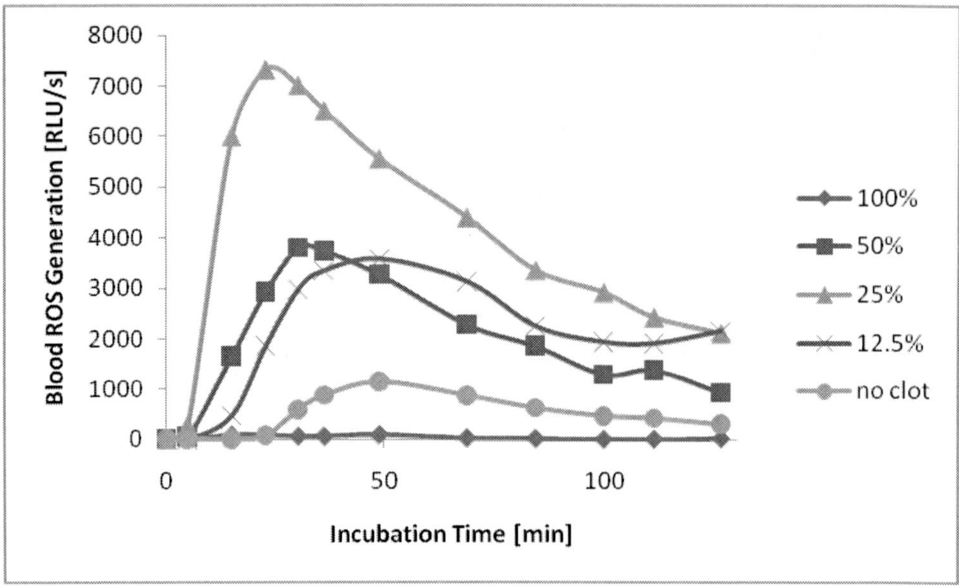

Figure 3. ROS generation induced by fresh plasmatic micro-thrombi in fresh EDTA-blood. Fresh normal citrated platelet poor plasma was recalcified and diluted to 50%, 25%, 12.5%. Immediately thereafter 100 μl portions were pipetted into a black microtiter plate. After 60 min (37°C) the 2.4 μg/ml zymosan A stimulated ROS generation in fresh EDTA-blood was determined. The second increase of ROS generation in 12.5% clots is due to blood micro-thrombi by recalcification during the BRGA [14].

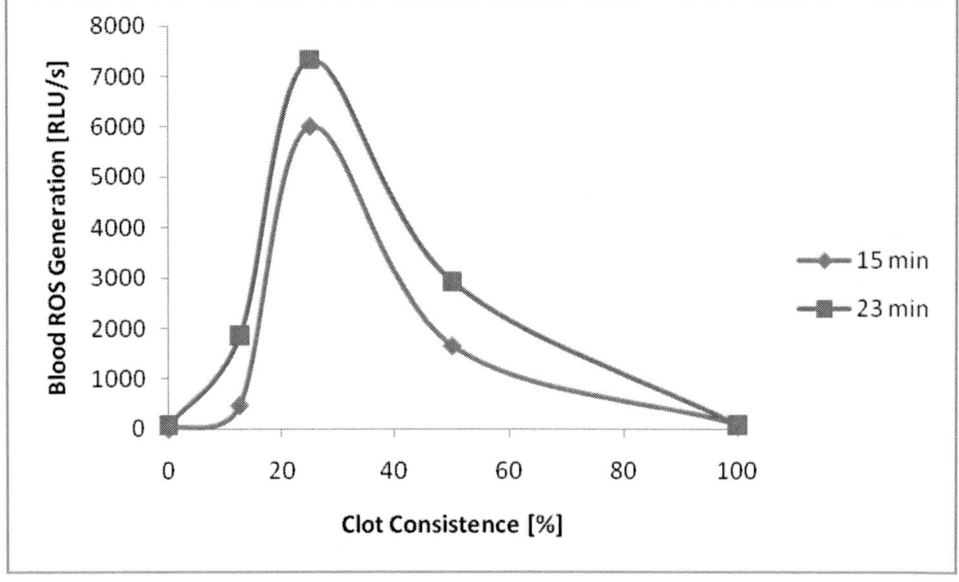

Figure 4. ROS generation induced by fresh plasmatic micro-thrombi in fresh EDTA-blood. Fresh normal citrated platelet poor plasma was recalcified and diluted to 50%, 25%, 12.5%. Immediately thereafter 100 μl portions were pipetted into a black polystyrene microtiter plate (Brand®781608). After 60 min (37°C) the 2.4 μg/ml zymosan A stimulated ROS generation in fresh EDTA-blood was determined at 15 and 23 min as described in Methods.

REFERENCES

[1]   Collen D, Lijnen HR. The fibrinolytic system in man. *Crit Rev Oncol Hematol.* 1986; 4: 249-301.

[2]   van Giezen JJ, Chung-A-Hing JE, Vegter CB, Bouma BN, Jansen JW. Fibrinolytic activity in blood is distributed over a cellular and the plasma fraction which can be modulated separately. *Thromb Haemost.* 1994; 72: 887-92.

[3]   Moir E, Robbie LA, Bennett B, Booth NA. Polymorphonuclear leucocytes have two opposing roles in fibrinolysis. *Thromb Haemost.* 2002; 87: 1006-10.

[4]   Stief TW. Neutrophil granulocytes in hemostasis. *Hemostasis Laboratory* 2008; 1: 269-89.

[5]   Wright SD, Weitz JI, Huang AJ, Levin SM, Silverstein SC, Loike JD. Complement receptor type three (CD11b/CD18) of human polymorphonuclear leukocytes recognizes fibrinogen. *Proc Natl Acad Sci U S A.* 1988; 85: 7734-8.

[6]   Nathan C. Neutrophils and immunity: challenges and opportunities. *Nat Rev Immunol.* 2006; 6: 173-82.

[7]   Stief TW. The blood fibrinolysis / deep-sea analogy: a hypothesis on the cell signals singlet oxygen/photons as natural antithrombotics. *Thromb Res* 2000; 99: 1-20.

[8]   Stief TW, Fareed J. The antithrombotic factor singlet oxygen/light ($^1O_2$/hv). *Clin Appl Thrombosis/Hemostasis* 2000; 6: 22-30.

[9]   Stief TW, Fu K, Yang LH, Ramaswamy A, Fareed J. Singlet oxygen ($^1O_2$) induces selective thrombolysis in vivo by massive granulocyte infiltration into the thrombus. 43. GTH Congress, Mannheim, 24.-27.2.1999, *Ann. Hematol.* 1999; 78: A32 (FV112).

[10]   Stief TW. The physiology and pharmacology of singlet oxygen. *Med Hypoth* 2003; 60: 567-572.

[11]   Stief TW. Regulation of hemostasis by singlet oxygen ($^1\Delta O_2$). *Curr Vasc Pharmacol* 2004; 2: 357-362.

[12]   Stief T. The routine blood ROS generation assay (BRGA) triggered by typical septic concentrations of zymosan A. *Hemostasis Laboratory* 2013; 6: 89-98.

[13]   Stief T. Glucose initially inhibits and later stimulates blood ROS generation. *Journal of Diabetes Mellitus* 2013, 3: 15-21.

[14]   Stief T. Micro-thrombi stimulate blood ROS generation. *Hemostasis Laboratory* 2013; 6: 315-25.

[15]   Haverkate F, Timan G. Protective effect of calcium in the plasmin degradation of fibrinogen and fibrin fragments D. *Thromb Res.* 1977; 10: 803-12.

[16]   Stief TW, Lenz P, Becker U. A simple method for producing degradation products of fibrinogen by an insoluble derivative of plasmin. *Thromb Res.* 1987; 48: 603-9.

[17]   Suenson E, Thorsen S. Secondary-site binding of Glu-plasmin, Lys-plasmin and miniplasmin to fibrin. *Biochem J.* 1981; 197: 619-28.

[18]   Bosma PJ, Rijken DC, Nieuwenhuizen W. Binding of tissue-type plasminogen activator to fibrinogen fragments. *Eur J Biochem.* 1988; 172: 399-404.

[19]   Stief TW, Marx R, Heimburger N. Oxidized fibrin(ogen) derivatives enhance the activity of tissue type plasminogen activator. *Thromb Res.* 1989; 56: 221-8.

[20] Suenson E, Bjerrum P, Holm A, Lind B, Meldal M, Selmer J, Petersen LC. The role of fragment X polymers in the fibrin enhancement of tissue plasminogen activator-catalyzed plasmin formation. *J Biol Chem.* 1990; 265: 22228-37.

[21] Stief TW, Kretschmer V, Kosche B, Doss MO, Renz H. Thrombin converts singlet oxygen ($^1O_2$)-oxidized fibrinogen into a soluble t-PA cofactor. A new method for preparing a stimulator for functional t-PA assays. *Ann Hematol.* 2001; 80: 189-94.

[22] Perera C, McNeil HP, Geczy CL. S100 Calgranulins in inflammatory arthritis. *Immunol Cell Biol.* 2010; 88: 41-9.

[23] Stief T. The true PAP concentration in plasma. *Hemost Lab.* 2014; 7 (issue 3).

[24] Stief TW. Oxidized fibrin stimulates the activation of pro-urokinase and is the preferential substrate of human plasmin. *Blood Coagul Fibrinolysis* 1993; 4: 117-22.

[25] Stief TW. The fibrinogen functional turbidimetric assay. *Clin Appl Thromb/Hemost.* 2008: 14: 84-96.

Chapter 2

SYSTEMIC INTRAVASCULAR COAGULATION

ABSTRACT

The rapid laboratory diagnosis of the pre-phase of pathologic (disseminated) intravascular coagulation (PIC phase 0) is a great clinical challenge. Typical PIC-1 means multiple micro-thrombi that obstruct vital organ areas, and that can cause multi-organ failure. There is clinical need for reliable hemostasis parameters to diagnose PIC-0 or PIC-1 to start the adequate treatment in time. The routine hemostasis tests such as prothrombin time and activated partial thromboplastin time are too blunted; furthermore, thrombocytes might be dysfunctional, cell fragments might be counted as platelets, platelets-count or fibrinogen concentration are acute phase parameters that might increase instead of decrease, and antithrombin-3 activity might change only slightly or it might be altered in an artificially diluted test matrix.

In the present review new routine parameters for diagnosis of the early phases of PIC are presented: basal chromogenic thrombin activity (systemic F2a), thrombin generation after plasma recalcification (RECA), F2a generation after intrinsic activation of plasma (INCA), or F2a generation after extrinsic activation of plasma (EXCA), fibrinogen function+antigen+ratio, undiluted antithrombin-3 activity, and active endotoxin (here clinically understood as active lipopolysaccharide or active ß-glucan). A combination of these new assays allows to diagnose the pre-phase of PIC within minutes. Especially the functional ultra-specific systemic F2a - Test (normal range 100±20%; 100%=5.5 mIU/ml) or the therefrom derived RECA helps to distinguish the normal systemic intravascular coagulation (NIC) from PIC-0 (121-150% of normal basal F2a) or from the PIC-1 (151-200% of normal basal F2a). All these simple photometric assays do not require special hemostasis machines to test the dynamic coagulation state and to differenciate between LPS- and ß-glucan- sepsis of each individual patient especially during the *"first six golden hours"* of the disease.

Keywords: Pathologic intravascular coagulation (PIC), normal intravascular coagulation (NIC), thrombin, endotoxin, LPS, ß-glucan

INTRODUCTION

In normal human blood there is always a basal level of hemostasis activation, that results in the generation of small amounts of thrombin. Systemic thrombin can be dangerous, because it converts fibrinogen into fibrin that might occlude blood vessels, the resulting micro-thrombi or thrombi cause ischemia/infarction of organs [1-3, 42]. Thrombin is therefore physiologically neutralized by antithrombins: antithrombin-3 (= AT-3 = the main inhibitor of thrombin), heparin cofactor-2, antithrombin-1 (= fibrinogen [4]), or α2-macroglobulin. Healthy normal plasma has 5.5±1.1 mIU/ml amidolytically active thrombin (F2a) that is mainly entrapped in and transported by α2-macroglobulin [5]. If hemostasis is pathologically activated by intrinsic or extrinsic triggers, pathologically elevated F2a activities arise [6]. Over-activation of intravascular coagulation is named pathologic (disseminated) intravascular coagulation (PIC); the most common causes are trauma, surgery, ischemia, liver disease, or sepsis [7, 8]. Intrinsic triggers are e.g. cell fragments such as membrane vesicles or DNA, the extrinsic trigger tissue factor (TF) is released upon damage of cells in TF-rich tissue, e.g. brain or placenta. Pathologic activities of systemic (non-local) thrombin generate multiple fibrin-thrombi in the blood circulation, which might cause multi-organ failure.

PIC can be acute or chronic; in acute PIC a large amount of thrombin activity in a short time period is generated, in chronic PIC small amounts of coagulation triggers result in increased but still compensated levels of thrombin activity [9]. The PIC phases are:

(0) hyper-coagulable pre-phase, often without thrombotic complications
(1) hyper-coagulable phase, with multiple micro-thrombi
(2) hyper/hypo-coagulable phase, consumption of clotting factors, eventually with bleeding ("overt DIC")
(3) hyper-fibrinolysis, with severe bleeding

Of great clinical importance are the phases (0) and (1), since these phases threaten the life of most patients (1). PIC is relatively easy to diagnose in phases (2) and (3), as abnormalities occur in global coagulation tests, such as the prothrombin time [9-11]. The diagnosis of phase (1) of PIC, and especially the diagnosis of phase (0) of PIC (the pre-phase of PIC) is an analytical challenge, because the usual laboratory parameters remain normal in very early PIC [7,9]; "non-overt DIC" (chronic, compensated) remains often undetected [8]. There is clinical need for reliable hemostasis parameters that diagnose the pre-phase of PIC (phase 0) (= PIC-0) to start the adequate treatment in time [7, 12].

HITHERTO ROUTINE HEMOSTASIS TESTS FOR PIC DIAGNOSIS

Prothrombin time (PT) and activated partial thromboplastin time (APTT) are blunted assays. They are of interest in hypo-coagulation, as it occurs e.g. in anticoagulant treatment by coumarins or unfractionated heparin, respectively. These tests do not detect spurious tendencies of hyper-coagulation, as it is typical for the very early phase of PIC. For PIC diagnosis, the traditional diagnostic trias of 1) decreasing platelet count, 2) decreasing antithrombin III activity, and 3) decreasing fibrinogen activity is used. However, 1) platelets might be dysfunctional, cell fragments (e.g. microparticles) might be counted as platelets, or the platelet number might increase in inflammation (partial acute phase reactant), 2) fibrinogen is an acute phase reactant that might increase in inflammation, and 3) antithrombin-3 activity changes only slightly or it might be altered in an artificially diluted test matrix.

D-dimer concentrations depend on the activities of

- thrombin,
- factor 13a, and
- plasmin,

i.e. D-dimer assays do not distinguish between activation of coagulation as in PIC and activation of fibrinolysis as in (primary) hyper-fibrinolysis. Besides, D-dimer may not increase in PIC [13]. Furthermore, the usual diagnosis of hemostasis activation is performed in citrated plasma, which is rather unstable: critically ill patients have pathologic activities of proteases in their blood [14] that might artificially modulate parameter concentrations, such as D-dimer, in the pre-analytic time interval.

NEW ROUTINE HEMOSTASIS TESTS FOR DIAGNOSIS OF EARLY PIC

Basal chromogenic thrombin (F2a) activity and therefrom derived ultra-specific, ultra-sensitive F2a generation tests, fibrinogen function+antigen+ratio, undiluted antithrombin activity, and active endotoxin (here clinically understood as active lipopolysaccharide or active ß-glucan) are new simple rapid routine tests [43] to diagnose the early phases of PIC within the first *six golden hours* [44].

SYSTEMIC F2A -TEST

Reliable data on *in vivo* thrombin generation in the blood of a critically ill patient are of great clinical interest to intervene in time in any pathologic disturbance of hemostasis. Previously, thrombin-antithrombin-complex (TAT) has been considered a marker for *in vivo* thrombin generation [15], however the TAT concentration depends also on the function of antithrombin-3 and on the function of the liver. A new marker for *in vivo* thrombin activity is the amidolytic thrombin activity in plasma: thrombin trapped in α_2-macroglobulin is chromogenically determined via cleavage of a tri-amino acids–paranitroanilide (pNA)-substrate [5]: EDTA blood of patients is centrifuged within 60 min (23°C), and the resulting plasma is immediately stabilized 1+1 with 2.5 M arginine, pH 8.6. In the F2a-Test one part of arginine-stabilized EDTA plasma can be incubated with one part of 0.8 mM fast chromogenic thrombin substrate (e.g. HD-CHG-Ala-Arg-pNA; MW: 624.6) in 1.25 M arginine, pH 8.7, preferably in half area microwells, and the increase in absorbance at 405 nm with time ($\Delta A/t$) at 37°C is determined by a microtitre plate reader with a 1 mA resolution (e.g. PHOmo;

Autobio-anthos, Krefeld, Germany) (Table 1). The F2a-Test is standardized against pooled normal EDTA-plasma (100 % of normal), also stabilized 1+1 with 2.5 M arginine, pH 8.6. Arginine-stabilization of EDTA-plasma conserves the hemostasis activation state of plasma even against freezing/thawing [14].

With the F2a-Test, the basal coagulation activation in normal blood can be measured, which reflects the normal intravascular coagulation (NIC). The normal range for basal F2a activity is 100±20 % (mean value (MV) ± 1 standard deviation (SD); 100 % is equivalent to 5.5 mIU/ml thrombin, that is 2.17 ng/ml active F2a; 1 mg F2a = 2525 IU F2a). In patients with severe bacterial or fungal sepsis the basal F2a activity increased to 173±62 % of normal (MV±1SD) [14]. Therefore, the expression PIC (derived from "coagulation pathology") seems better than "disseminated intravascular coagulation". The F2a-Test is especially useful for PIC-0 or PIC-1. In PIC-2 or PIC-3 the basal F2a activity might paradoxically decrease a bit due to partial destruction of α2-macroglobulin by high activities of aggressive proteases in the blood of a severely ill patient. Therefore, for a patient in late PIC-2 or in PIC-3 an abnormal PT adds further information on the extreme illness.

The F2a-Test is ultra-specific and ultra-sensitive for thrombin. Supra-molar concentrations of arginine completely inhibit disturbing kallikrein and depolymerize uncrosslinked fibrin. Blood of patients with severe pancreatitis might contain in the range of about 100 ng/ml active trypsin. In presence of supra-molar concentrations of arginine, 100 ng/ml active trypsin acts in the F2a-Test only like about 2 ng/ml active thrombin (= approximately 100 % of normal basal thrombin in plasma). Since trypsin activates the intrinsic phase of coagulation [16], a strongly positive result of the F2a-Test in severe pancreatitis really reflects a clinical prothrombotic state.

Table 1. Exemplary F2a-Test

50 µl EDTA-1.25 M arginine, pH 8.7 plasma 50 µl 0 or 0.8 mM HD-CHG-Ala-Arg-pNA in 1.25 M arginine
$\Delta A_{405nm}/t$ (37°C)

ULTRA-SPECIFIC F2A GENERATION TESTS

Derived from the above described F2a-Test, 3 ultra-specific, ultra-sensitive chromogenic tests of F2a generation have been developed [6]:

1. RECA (recalcified coagulation activity assay [17])
2. INCA (intrinsic coagulation activity assay [18])
3. EXCA (extrinsic coagulation activity assay [19]).

In all these tests, 10 parts of fresh citrated plasma are incubated with 1 part of coagulation trigger, that is 250 mM $CaCl_2$ in the RECA, ellagic acid + 250 mM $CaCl_2$ in the INCA, and 1 ng/ml tissue factor in 5 % human albumin + 250 mM $CaCl_2$ in the EXCA. In the then following coagulation reaction time (CRT; 37°C) thrombin is generated. The CRT for RECA is about 15-20 min (main value) and 25 min (control value). For INCA the CRT is 4 min (main value) and 5 min (control value). For EXCA the CRT is 1 min (main value) and 2 min (control value). The assay results are only valid, if the ratio: (F2a in main value)/ (F2a in control value) is less than 1, i.e. if the thrombin generation kinetic is in its ascending part (thrombin is not significantly trapped within antithrombin I (= fibrin [4])). The test results can be calibrated against purified thrombin in 6.7 % human albumin, against pooled normal plasma, or against the mean value of unfrozen individual normal plasmas. The approximate normal ranges of these new thrombin generation tests are 100%±20-30% of normal; 100 % of normal = about 0.15 IU/ml F2a in the RECA, depending on the nature of the microwells, untreated polystyrene being best suited (e.g. U-wells 701300 or 781600 from Brand, Wertheim, Germany), 100 % of normal = about 0.5 IU/ml F2a in the INCA or in the EXCA. The RECA diagnoses the pro-thrombotic capacity of citrated plasma [6], if it is performed under well standardized pre-analytic conditions, i.e. immediate transport of citrated blood sample at 23±2°C to the central laboratory, immediate centrifugation at 2800 g for 5 min (23°C), and immediate analysis of plasma.

FIBRINOGEN FUNCTION+ANTIGEN+RATIO

Normal fibrinogen activity is important for normal hemostasis function. Too much fibrinogen function (> 140 % of normal) might be pro-thrombotic, too few fibrinogen function (< 60 % of normal) might predispose to cerebral hemorrhage. Fibrinogen function is usually measured by addition of unphysiologically high activities of thrombin (> 10 IU/ml final assay concentration). Two new fibrinogen tests have been described, the FIFTA (fibrinogen functional turbidimetric assay) (Table 2a) [20] and the FIATA (fibrinogen antigen turbidimetric assay) (Table 2b) [21]. In contrast to hitherto fibrinogen tests, the FIFTA works with a final physiologic thrombin activity of only 0.2 IU/ml, i.e. with the approximate thrombin activity in physiologically clotting normal plasma [17,22]. Often commercial functional fibrinogen tests depend on the plasma matrix: pathologically low albumin concentrations change the turbidimetric fibrin signal: this can result in a false fibrinogen concentration in patients with severe changes of the plasma matrix (plasmas of critically ill patients in intensive care units often have albumin concentrations of only 10 20 g/l (normal MV±1SD= 44±4g/l)). In the FIFTA, 1 part of citrated plasma is incubated with 2 parts of FIFTA-reagent, containing 0.3 IU/ml thrombin, 0.4 mg/ml polybrene®, 6 % human albumin, phosphate buffered saline (PBS). The turbidity increase at 405 nm within 2 min (37°C) or within 5 min (23°C) is measured.

In the FIATA, fibrinogen and fibrinogen-like molecules are precipitated by the antibiotic vancomycin: 1 part of plasma is incubated with 2 parts of PBS, the turbidity at 405 nm is determined, 2 parts of 4.4 mM vancomycin in PBS are added and the turbidity increase is measured after 2 min (23°C) or 40 s (37°C).

FIFTA and FIATA are standardized with pooled normal citrated plasma (=100 % of normal). Via FIATA the fibrinogen concentration can be determined even in EDTA-plasma. The FIFTA/FIATA ratio (FI ratio) indicates the presence of normo-reactive (ratio=1.0±0.1), over-reactive (FI ratio > 1.1), or under-reactive (FI ratio < 0.9) fibrinogen. Over-reactive fibrinogen might occur in presence of enhancers of fibrin-polymerization such as soluble fibrin; under-reactive fibrinogen might occur in presence of dysfibrinogens or fibrinogen degradation products by plasmin.

The normal range for FIFTA is 100±20%, septic patients had 188±66% of normal, the normal range for FIATA is 100±20%, septic patients had 179±66% of normal, and the normal range of FIFTA/FIATA ratio is 1.0±0.1, septic patients had 1.12±0.32.

Table 2a. Exemplary FIFTA (fibrinogen function)

| 20 µl citrated plasma |
| 40 µl 0.3 IU/ml thrombin, 0.4 mg/ml polybrene®, 6 % human albumin, PBS |
| ΔA_{405nm}/2 min (37°C) |

Table 2b. Exemülary FIATA (fibrinogen antigen)

| 10 µl plasma |
| 20 µl PBS |
| A_{405nm} |
| 20 µl 4.4 mM vancomycin in PBS |
| ΔA_{405nm}/2 min (23°C) |

UNDILUTED ANTITHROMBIN-3 ACTIVITY (AT3)

AT3 is the most important physiological inhibitor of thrombin [23-25]. Decreased activity of AT3 results in increased thrombin generation. Therefore, AT3 activity must be determined routinely. Usual AT3 determinations work with a > 50fold dilution of plasma to overcome turbidity problems within the assay nascending fibrin [26-28]. A new AT3 test has been developed, that works with nearly undiluted plasma [29]: in the thrombin reaction phase, 10 µl plasma or standards (100 % of normal pooled plasma and this plasma diluted with 0.9 % NaCl) are incubated with 2 µl thrombin reagent = 58 IU/ml (9.7 IU/ml final) thrombin, 6 % human albumin, 54 IU/ml (9 IU/ml final) unfractionated heparin, 1600 mM (267 mM final) arginine, pH 7.4, for 30s (23°C) in microwells with flat bottom (F-wells). Then 50 µl 2.5 M arginine, pH 8.6, 0.124 % Triton X 100® are added. After 3 min (23°C), the thrombin detection phase starts with addition of 20 µl 1 mM (0.24 mM final) chromogenic thrombin substrate HD-CHG-Ala-Arg-pNA in 1.25 M (1.87 M final) arginine, pH 8.7. ΔA/t at 405 nm is determined (Table 3). The normal range for this new AT3 test is 100±15% (MV±1SD), septic patients have 47±27% of normal active AT3. Improving AT3 activity correlates with the survival of the patient.

ACTIVE ENDOTOXIN = ENDOTOXIN REACTIVITY

Endotoxin is here clinically understood as lipopolysaccharide (LPS) from gram-negative bacteria or ß-glucan from fungi. The human organism is especially susceptible to active endotoxin [30]. Active (=free, unbound) endotoxin destroys susceptible cells in the blood stream (e.g. monocytes), which triggers intrinsic coagulation. Kallikrein and thrombin are formed, and a PIC may arise. Therefore, it is of great clinical importance to measure active endotoxin routinely to diagnose a possible pre-phase of sepsis and of PIC as soon as possible to start the adequate treatment in time. Active endotoxin can be measured in plasma within minutes by coagulation factors of the *Limulus* cascade: 10 µl plasma (with EDTA or citrate) or plasma-standards (1, 10, 100 ng/ml LPS reactivity) are incubated for 10 min (37°C) with 60 mM chloramine-T® in 50 mM sodium citrate, pH 7.4. 10 µl 230 mM methionine are added. After 1 min (37°C) 30 µl *Limulus*-reagent (*Limulus* factors + chromogenic substrate Ile-Glu-Gly-Arg-pNA) are added and the increase in absorbance with time ($\Delta A/t$) is measured (Table 4a).

If only active ß-glucan is to be quantified, the reaction conditions are the same as for active endotoxin, with the exception that the methionine reagent contains also 0.63 mg/ml of the LPS-inhibitor polymyxin B, which inhibits LPS reactivities of up to 1000 ng/ml LPS added to normal plasma (Table 4b).

Table 3. Exemplary undiluted AT3 activity test

10 µl plasma 2 µl 58 IU/ml thrombin, 6 % human albumin, 54 IU/ml heparin, 1600 mM arginine, pH 7.4
30s (23°C)
50 µl 2.5 M arginine, pH 8.6, 0.124 % TritonX100® 20 µl 1 mM HD-CHG-Ala-Arg-pNA, 1.25 M arginine, pH 8.7
$\Delta A_{405nm}/t$

Table 4a. Exemplary active endotoxin (here: active LPS + active ß-glucan) test

10 µl plasma 10 µl 60 mM chloramine-T® in 50 mM sodium citrate, pH 7.4
10 min (37°C)
10 µl 230 mM methionine
1 min (37°C)
30 µl *Limulus*-reagent (*Limulus* factors + pNA-substrate, pH 8)
0, 10, 30, 60 min (37°C)
ΔA_{405nm}

Table 4b. Exemplary active ß-glucan test

10 µl plasma 10 µl 60 mM chloramine-T® in 50 mM sodium citrate, pH 7.4
10 min (37°C)
10 µl 230 mM methionine, 0.63 mg/ml (4957 IU/ml) polymyxin B
1 min (37°C)
30 µl *Limulus*-reagent (*Limulus* factors + pNA-substrate, pH 8)
0, 10, 30, 60 min (37°C)
ΔA_{405nm}

Active LPS (LPS reactivity) is calculated with the formula:

LPS reactivity [mA/t] = endotoxin reactivity [mA/t] minus ß-glucan reactivity [mA/t]

Reactivity examples: 10 ng/ml LPS reactivity corresponds to the LPS reactivity of normal plasma (containing 0.8 ng/ml LPS) that had been supplemented with additional 9.2 ng/ml purified LPS. 10 µg/ml ß-glucan reactivity corresponds to the ß-glucan reactivity of normal plasma (containing < 0.1 µg/ml ß-glucan) that had been supplemented with 10 µg/ml purified (1→3)-ß-D-glucan.

The normal range of LPS reactivity is 100±25 % (100% = 0.8 ng/ml). Pre-septic patients have about 2 ng/ml (200-300 % of normal) LPS reactivity. The LPS reactivity in patients with severe sepsis can increase up to about 100 ng/ml (mean value=20.5 ng/ml, standard deviation=19.6 ng/ml) [14,34].

The fungal pathology should never be underestimated: massive fungal infection or a combined bacterial/fungal sepsis can cause PIC [31,32]. Normal human plasma has < 0.1 µg/ml ß-glucan reactiviy.

Pre-septic patients have about 0.5 µg/ml ß-glucan reactivity. The ß-glucan reactivity in patients with severe sepsis can increase up to about 50 µg/ml (mean value = 4.7 µg/ml, standard deviation =7.0 µg/ml). Since free LPS is about 300fold as toxic as free ß-glucan [6,14], the load of active endotoxin of a patient can be calculated as: LPS [ng/ml] + ß-glucan [ng/ml]/300. Then patients with severe sepsis had 36.2±26.3 ng/ml active endotoxin, of these pathological endotoxin reactivities about one third had LPS, one third hat ß-glucan, one third had LPS+ß-glucan [14]. In normal plasma, only about 0.01 % of LPS or about 20 % of ß-glucan are active=unbound=free [46].

SPECIAL ANTIGENTIC PARAMETERS FOR PIC DIAGNOSIS

Soluble Intercellular adhesion molecule-1 (sICAM-1) is a parameter for endothelial damage (normal range: 0.19±0.04 µg/ml, septic patients: 2.56±2.48 µg/ml) [14]. Plasmin-antiplasmin-complex (PAP; normal range: 100±30%, septic patients: 313±307%) and D-dimer (normal range: 0.05±0.03 µg/ml, septic patients: 4.0±3.6 µg/ml) indicate fibrinolysis activation [14]. Since D-dimer is a final product of the combined action thrombin, factor 13a, and plasmin, D-dimer tests cannot distinguish between coagulation activation and fibrinolysis activation.

To obtain reliable test results for unstable blood, these antigenic parameters should be measured in 1.25 M arginine - stabilized EDTA-plasma [14]. The inhibited plasmatic fibrinolysis in severe sepsis can be monitored by determination of plasminogen activator inhibitor-1 (PAI-1) activity or of the global assay FIPA (fibrinolysis parameters assay with 10 min reaction time) that measures the global plasmatic fibrinolytic state [33]: PAI-1 is about 10fold elevated and FIPA is about 5fold depressed in the begin of severe sepsis. An improvement of fibrinolysis seems to correlate with survival of the patient: the FIPA mean value approximately doubles within a stay of about 1 week in intensive care unit [34].

The dynamic hemostasis state of each individual patient should be determined routinely, as indicated in exemplary figures 1-5 [14, 30]. The low plasma volumes required for these new tests allows the determination of a complete state of hemostasis activation also in pediatrics [25,39-41]. The above reviewed new hemostasis assays help to diagnose the very early phase of PIC within minutes.

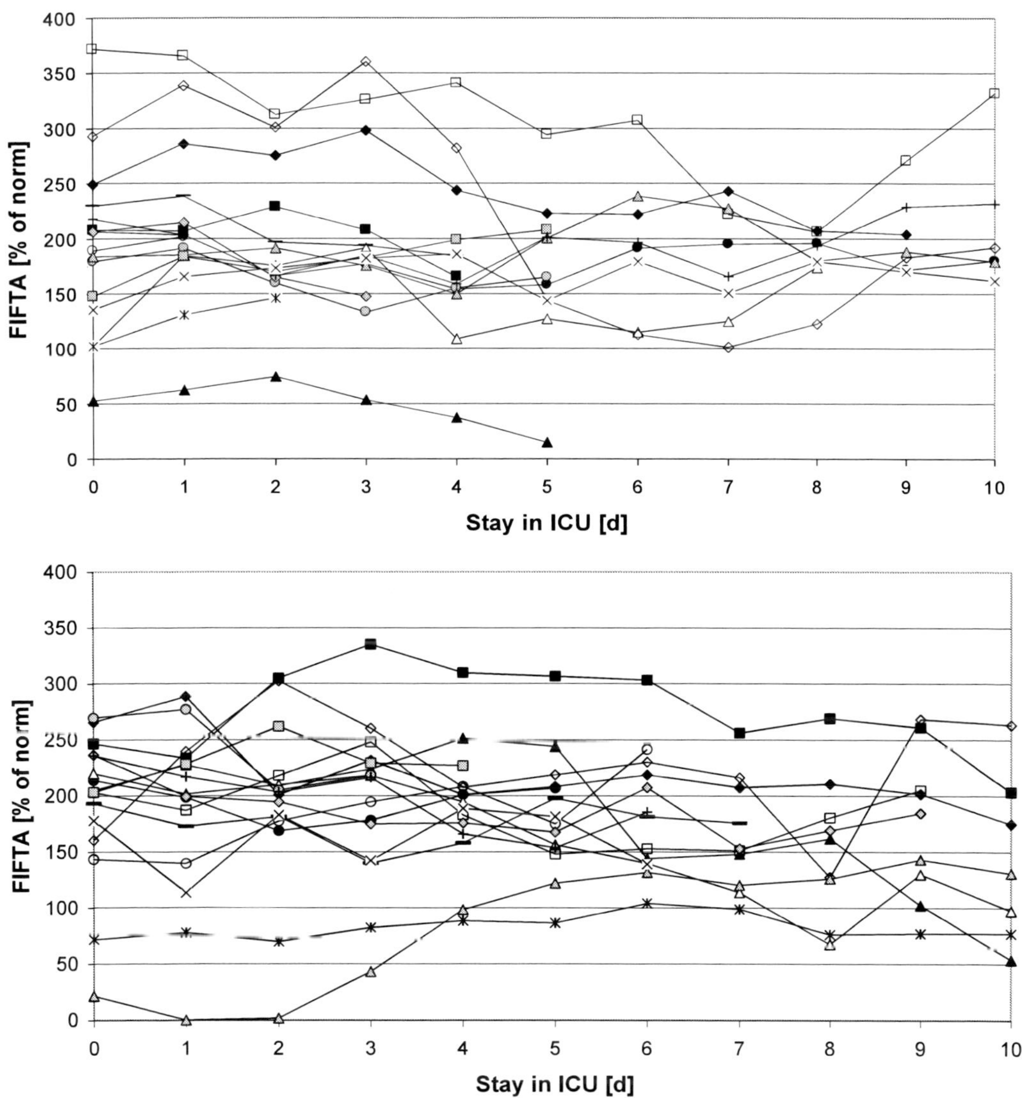

Figure 1. Fibrinogen function (FIFTA) in septic patients. Citrated plasma of 32 intensive care unit (ICU) patients with severe sepsis was analyzed for fibrinogen function with the FIFTA. Figure 1a= patients 1-16, Figure 1b= patients 17-32, each patient has his individual symbol. Patients with < 80 % FIFTA have pathologically decreased fibrinogen activities, e.g. by consumption of fibrinogen (PIC phase (2)) or by destruction of fibrinogen by plasmin (PIC phase (3)); patients with > 120 % FIFTA have pathologically increased fibrinogen activities, e.g. by hyper-fibrinogenemia or by enhancers of fibrin polymerization. Admission to ICU = stay of 0 days = day 1 in ICU.

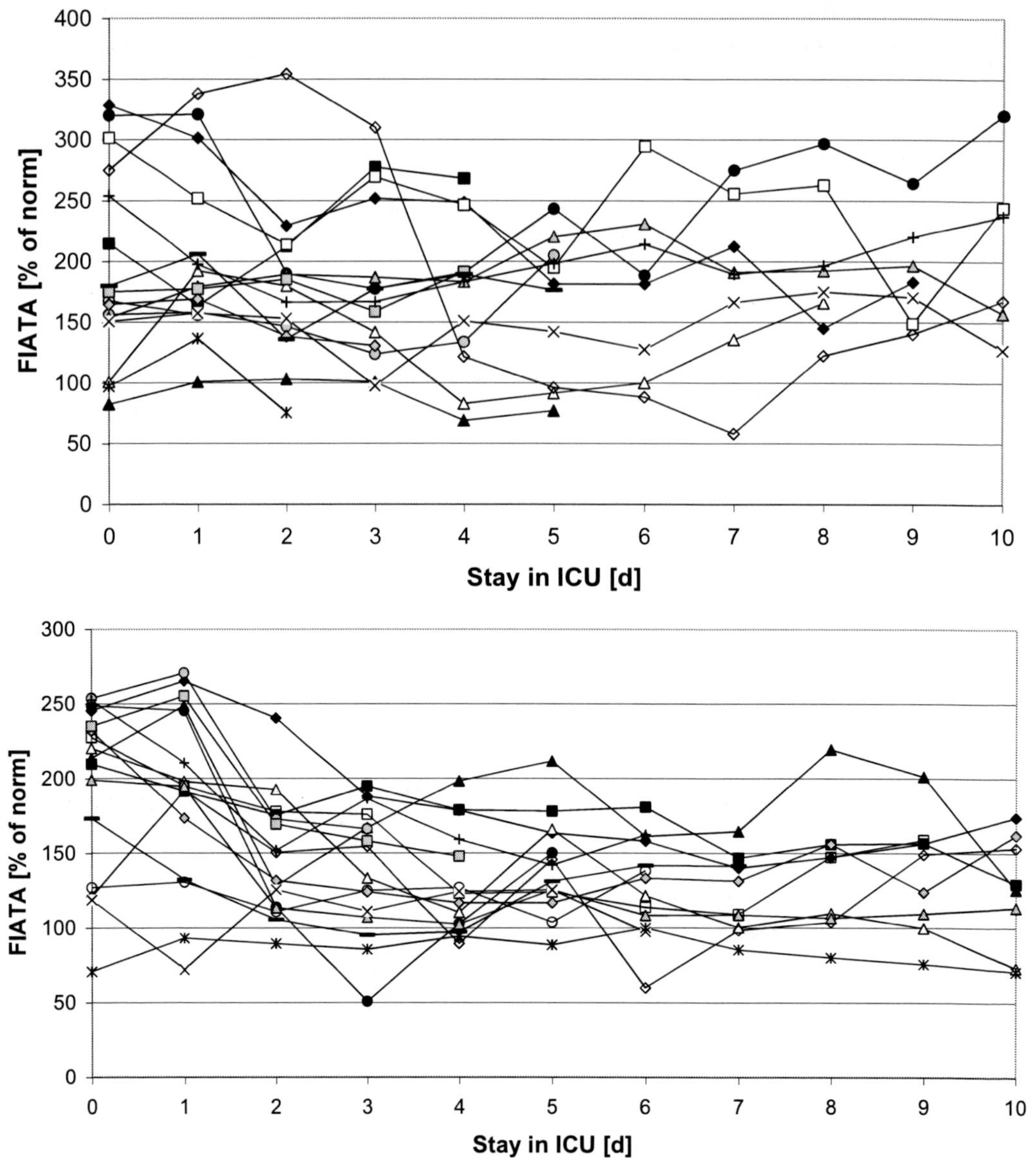

Figure 2. Fibrinogen antigen (FIATA) in septic patients. Citrated plasma of 32 intensive care patients with severe sepsis was analyzed for fibrinogen antigen by the FIFTA. Figure 2a= patients 1-16, Figure 2b= patients 17-32, each patient has his individual symbol.

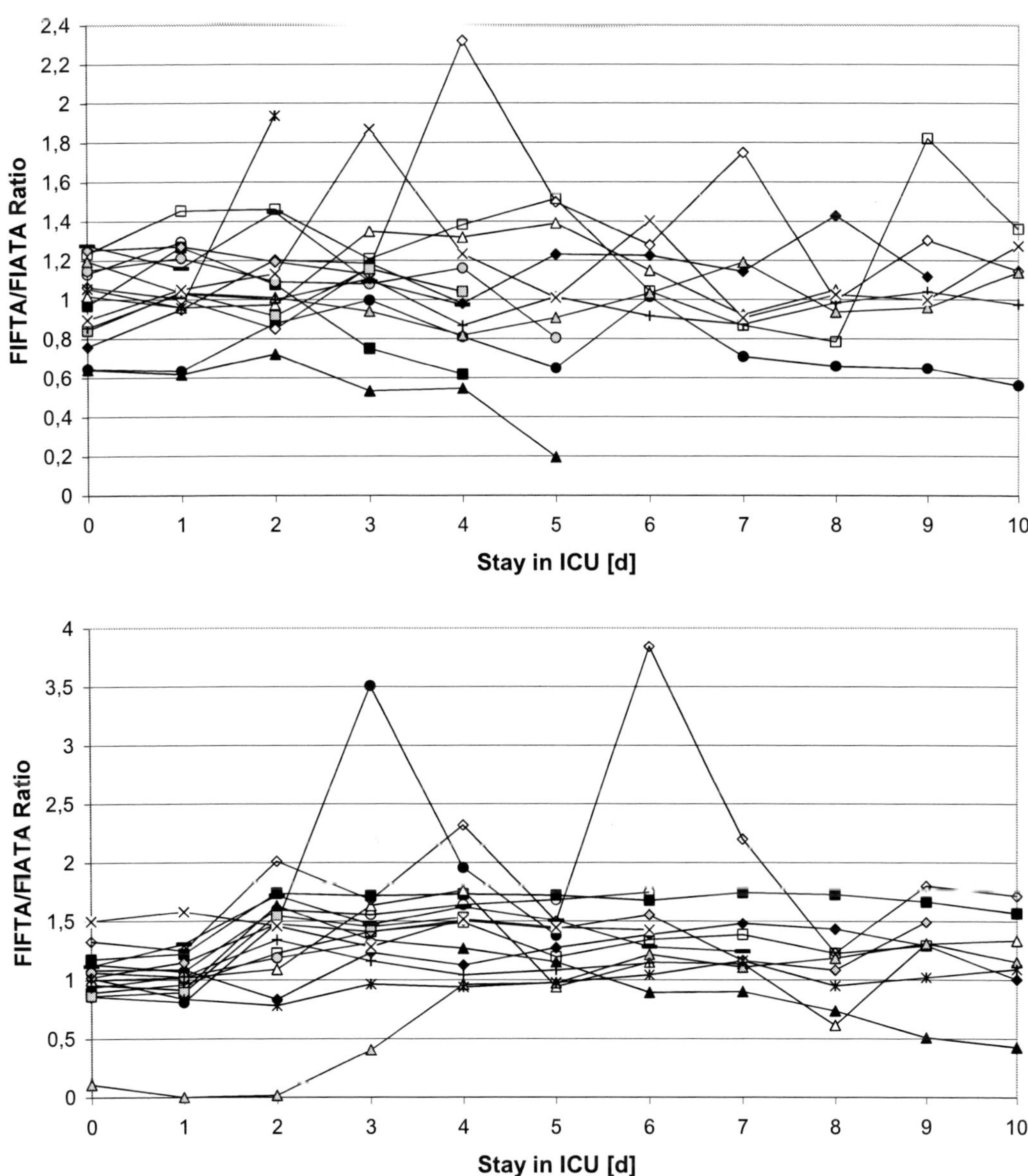

Figure 3. FIFTA/FIATA ratio in septic patients. The FIFTA/FIATA ratio (FI ratio) of the patient plasmas of Figure 1 and Figure 2 was calculated. Figure 3a= patients 1-16, Figure 3b= patients 17-32, each patient has his individual symbol. Patients with a FIFTA/FIATA ratio < 0.9 have under-reactive fibrinogen, e.g. fibrinogen degradation products as occurring in PIC phase (3) or dysfibrinogens; patients with a FI ratio > 1.1 have fibrinogen, whose function is over-reactive, e.g. due to enhancers of fibrin polymerization.

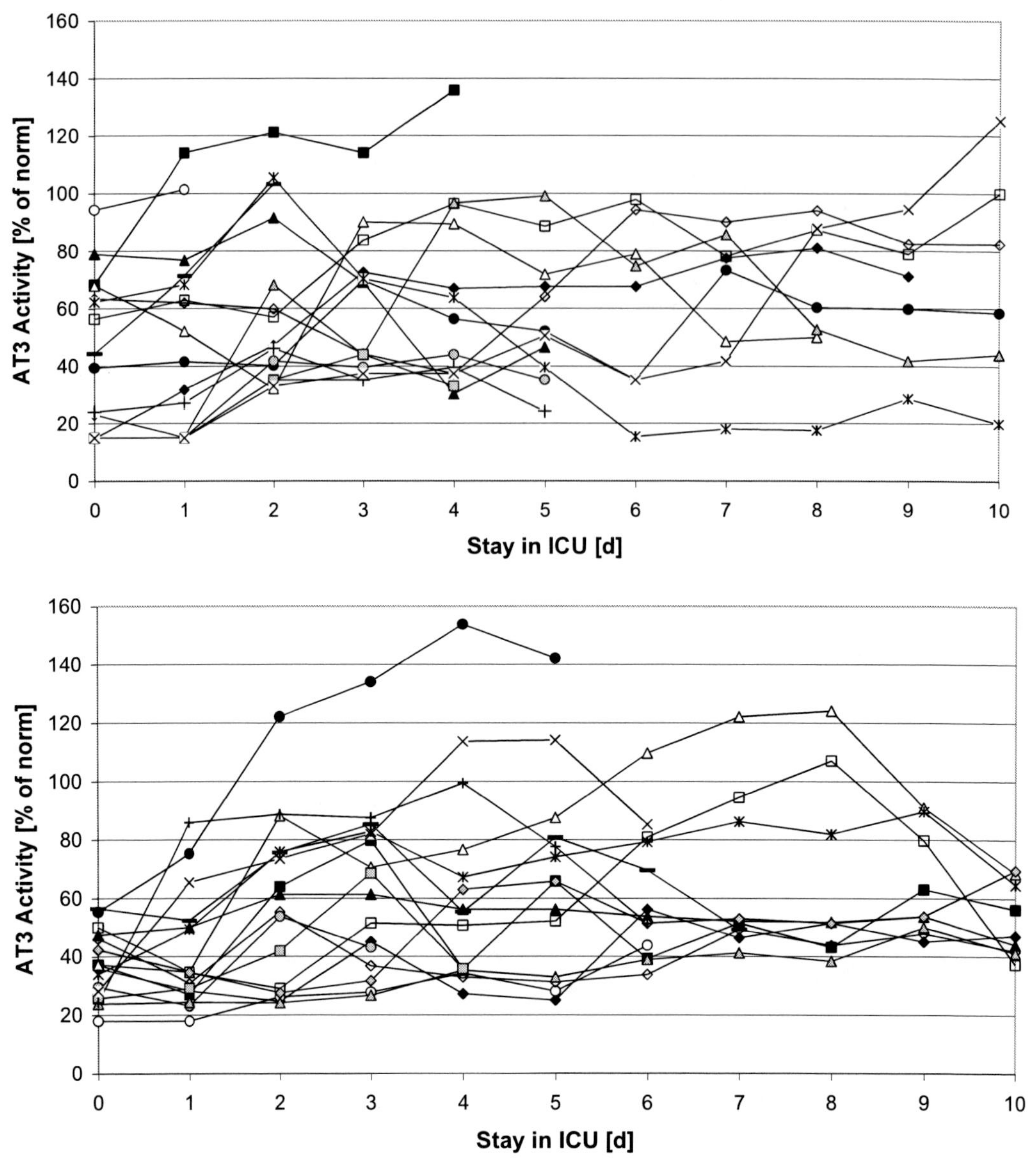

Figure 4. Undiluted antithrombin III function in septic patients. Citrated plasma of 32 intensive care patients with severe sepsis was analyzed for undiluted AT3 activity. Figure 4a= patients 1-16, Figure 4b= patients 17-32, each patient has his individual symbol. Normal range = 100 ±15 %.

The F2a-Test (normal range 100±20%; 100%= 5.5 mIU/ml basal F2a) or the RECA helps to distinguish the normal intravascular coagulation (NIC; 100±20 % basal F2a; F2a generation in RECA = 100 % ± 30 % of normal) from a pre-phase of PIC or from chronic PIC [36] where the basal F2a activity is 121-150% or the basal F2a activity is dynamically increasing, or the RECA values are > 130 % of normal. The blood half life of α2-macroglobulin/protease complexes is only 15-60 min, especially cells of the reticuloenthelial system possess such receptors [37, 38], which means that the α2-macroglobulin/thrombin concentration is not primarily dependent on the liver function.

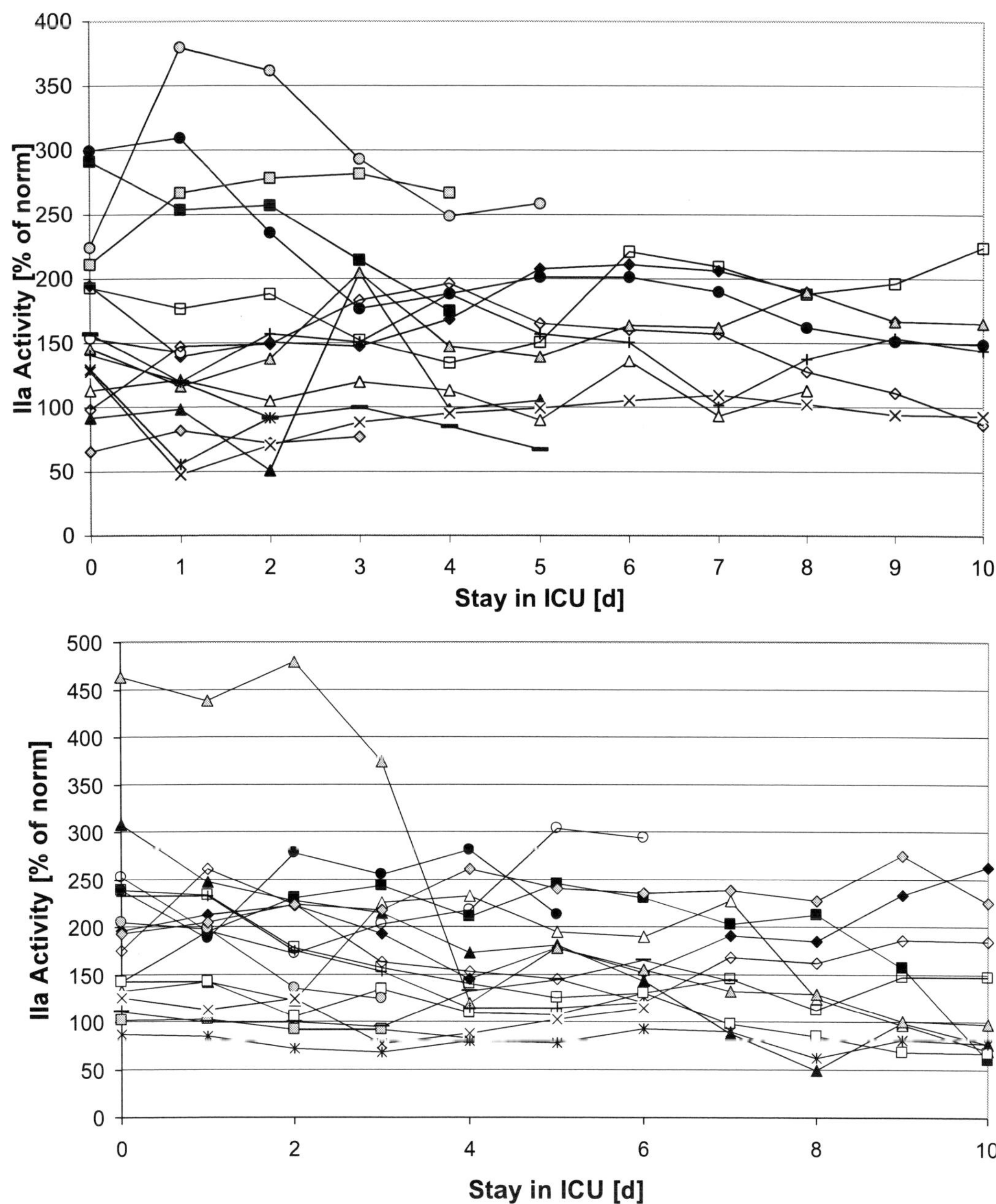

Figure 5. F2a-Test in septic patients. EDTA-1.25 M arginine plasma of 32 intensive care patients with severe sepsis was analyzed for basal thrombin (IIa) activity [14]. Figure 5a= patients 1-16, Figure 5b= patients 17-32, each patient has his individual symbol. Patient 26 (grey△) had pancreatitis, his F2a activities in Figure 5b (at ICU days 0-3) have to be 2fold multiplied F2a activities of 121-150% of norm and/or dynamically increasing F2a activities below 120 % of norm are characteristic for PIC phase (0). F2a activities of 151-200 % of norm and/or dynamically increasing F2a activities in the range 121-150 % of norm are characteristic for PIC phase (1).

CONCLUSION

PIC is comparable with a toxic snake bite [30,35] and very often results into severe prothrombotic hemostasis alterations with insufficient (frustrane) activation of cellular fibrinolysis [42].

Acute PIC is a terrible complication of a broad range of diseases; its high mortality is often caused by too late diagnosis and treatment [8], e.g. the first six hours of sepsis might be termed the "golden hours" when diagnosis and therapy is still in time for a good prognosis [43-46]. An aggressive disorder should be treated aggressively [1], LMWH is the first therapeutic option [47], if necessary a good AT3 drug should be combined [48].

REFERENCES

[1]   Bick RL. Disseminated intravascular coagulation current concepts of etiology, pathophysiology, diagnosis, and treatment. *Hematol Oncol Clin North Am* 2003; 17: 149-76.

[2]   Pasternak JJ, Hertzfeldt DN, Stanger SR, Walter KR, Werts TD, Marienau ME, Lanier WL. Disseminated intravascular coagulation after craniotomy. *J Neurosurg Anesthesiol* 2008; 20: 15-20.

[3]   Cotovio M, Monreal L, Navarro M, Segura D, Prada J, Alves A. Detection of fibrin deposits in horse tissues by immunohistochemistry. *J Vet Intern Med* 2007; 21: 1083-9.

[4]   Mosesson MW. Fibrinogen and fibrin structure and functions. J Thromb Haemost 2005; 3: 1894-904.

[5]   Stief TW. Specific determination of plasmatic thrombin activity. *Clin Appl Thromb Hemost* 2006; 12: 324-9.

[6]   Stief TW. Thrombin generation by exposure of blood to endotoxin: a simple model to study disseminated intravascular coagulation. *Clin Appl Thromb Hemost* 2006; 12: 137-61.

[7]   Bick RL. Disseminated intravascular coagulation. Objective laboratory diagnostic criteria and guidelines for management. *Clinics in Laboratory Medicine* 1994; 14: 729-69.

[8]   Urge J, Strojil J. Early diagnosis of DIC development into the overt phase. *Biomed Pap Med Fac Univ Palacky Olomouc Czech Repub* 2006; 150: 267-9.

[9]   Saba HI, Morelli GA. The pathogenesis and management of disseminated intravascular coagulation. *Clin Adv Hematol Oncol.* 2006; 4: 919-26.

[10]   Gando S, Saitoh D, Ogura H, Mayumi T, Koseki K, Ikeda T, Ishikura H, Iba T, Ueyama M, Eguchi Y, Ohtomo Y, Okamoto K, Kushimoto S, Endo S, Shimazaki S; Japanese Association for Acute Medicine Disseminated Intravascular Coagulation (JAAM DIC) Study Group. Natural history of disseminated intravascular coagulation diagnosed based on the newly established diagnostic criteria for critically ill patients: Results of a multicenter, prospective survey. *Crit Care Med* 2008; 36: 145-50.

[11]   Wada H, Hatada T. Pathophysiology and diagnostic criteria for disseminated intravascular coagulation associated with sepsis. *Crit Care Med* 2008; 36: 348-9.

[12]   Levi M. Disseminated intravascular coagulation. *Crit Care Med* 2007; 35: 2191-5.

[13] Constantinescu AA, Berendes PB, Levin MD. Disseminated intravascular coagulation and a negative D-dimer test. *Neth J Med* 2007; 65: 398-400.

[14] Stief TW, Ijagha O, Weiste B, Herzum I, Renz H, Max M. Analysis of hemostasis alterations in sepsis. *Blood Coagul Fibrinolysis* 2007; 18: 179-86.

[15] Bauer KA, Rosenberg RD. Thrombin generation in acute promyelocytic leukemia. *Blood* 1984; 64: 791-6.

[16] Stief TW. Quantification of thrombin generation by trypsin. *Medical and Biological Sciences (Scientific Journals International)* 2007; 1/ 2: 1-8.

[17] Stief TW. The Recalcified Coagulation Activity. *Clin Appl Thromb Hemost* 2008; 14: 369-78.

[18] Stief TW, Otto S, Renz H. The intrinsic coagulation activity assay. *Blood Coagul Fibrinolysis* 2006; 17: 369-78.

[19] Stief TW, Wieczerzak A, Renz H. The extrinsic coagulation activity assay. *Clin Appl Thromb Hemost* 2008; 14: 303-18.

[20] Stief TW. The fibrinogen functional turbidimetric assay. *Clin Appl Thromb Hemost* 2008; 14: 84-96.

[21] Stief TW. The fibrinogen antigen turbidimetric assay (FIATA). The $X^2\bar{X}$ test: the corrected chi-square comparison against the control-mean. *Clin Appl Thromb Hemost* 2007; 13: 73-100.

[22] Stief TW. Inhibition of thrombin generation in recalcified plasma. *Blood Coagul Fibrinolysis* 2007; 18: 751-60.

[23] Abildgaard U. Antithrombin--early prophecies and present challenges. *Thromb Haemost* 2007; 98: 97-104.

[24] Sorg H, Hoffmann JN, Rumbaut RE, Menger MD, Lindenblatt N, Vollmar B. Efficacy of antithrombin in the prevention of microvascular thrombosis during endotoxemia: An intravital microscopic study. *Thromb Res* 2007; 121: 241-8.

[25] St Peter SD, Little DC, Calkins CM, Holcomb GW 3rd, Snyder CL, Ostlie DJ. The initial experience of antithrombin III in the management of neonates with necrotizing enterocolitis. *J Pediatr Surg* 2007; 42: 704-8.

[26] Blomback M, Blomback B, Olsson P, Svendsen L. The assay of antithrombin using a synthetic chromogenic substrate for thrombin. *Thromb Res* 1974; 5: 621-32.

[27] Fareed J, Messmore HL, Walenga JM, Bermes EW, Bick RL. Laboratory evaluation of antithrombin III: a critical overview of currently available methods for antithrombin III measurements. *Semin Thromb Hemost* 1982; 8: 288-301.

[28] Gallimore MJ, Friberger P. Chromogenic peptide substrate assays and their clinical applications. *Blood Rev* 1991; 5: 117-27.

[29] Stief TW. Antithrombin III determination in nearly undiluted plasma. *Laboratory Medicine* 2008; 39: 46-8.

[30] Stief TW. Coagulation Activation by Lipopolysaccharides. *Clin Appl Thromb Hemost* 2007 Dec 26; [Epub ahead of print].

[31] Kalinski T, Jentsch-Ullrich K, Fill S, König B, Costa SD, Roessner A. Lethal *candida* sepsis associated with myeloperoxidase deficiency and pre-eclampsia. *APMIS* 2007; 115: 875-80.

[32] Lai CC, Liaw SJ, Lee LN, Hsiao CH, Yu CJ, Hsueh PR. Invasive pulmonary aspergillosis: high incidence of disseminated intravascular coagulation in fatal cases. *J Microbiol Immunol Infect* 2007; 40: 141-7.

[33] Stief TW, Fröhlich S, Renz H. Determination of the global fibrinolytic state. *Blood Coagul Fibrinolysis* 2007; 18: 479-87.

[34] Stief TW, Ulbricht K, Renz H, Max M. Plasmatic fibrinolysis in sepsis. *Hemostasis Laboratory* 2008; 1: 61-75.

[35] Paul V, Pudoor A, Earali J, John B, Anil Kumar CS, Anthony T. Trial of low molecular weight heparin in the treatment of viper bites. *J Assoc Physicians India* 2007; 55: 338-42.

[36] Iyoda M, Suzuki H, Ashikaga E, Nagai H, Kuroki A, Shibata T, Kitazawa K, Akizawa T. Elderly onset systemic lupus erythematosus (SLE) presenting with disseminated intravascular coagulation (DIC). *Clin Rheumatol* 2008; 27 Suppl 1: S15-8.

[37] Ohlsson K. Elimination of 125-I-trypsin alpha-macroglobulin complexes from blood by reticuloendothelial cells in dogs. *Acta Physiol Scand* 1971; 81: 269-72.

[38] Borth W. Alpha 2-macroglobulin, a multifunctional binding protein with targeting characteristics. *FASEB J* 1992; 6: 3345-53.

[39] Goldenberg NA, Manco-Johnson MJ. Pediatric hemostasis and use of plasma components. *Best Pract Res Clin Haematol* 2006; 19: 143-55.

[40] Kowalik MM, Smiatacz T, Hlebowicz M, Pajuro R, Trocha H. Coagulation, coma, and outcome in bacterial meningitis--an observational study of 38 adult cases. *J Infect* 2007; 55: 141-8. Epub 2007 Apr 2.

[41] Moloney-Harmon PA. Pediatric sepsis: the infection unto death. *Crit Care Nurs Clin North Am* 2005; 17: 417-29.

[42] Langer F, Spath B, Haubold K, Holstein K, Marx G, Wierecky J, Brümmendorf TH, Dierlamm J, Bokemeyer C, Eifrig B. Tissue factor procoagulant activity of plasma microparticles in patients with cancer-associated disseminated intravascular coagulation. *Ann Hematol* 2008; 87: 451-7.

[43] Stief TW. Innovative tests of plasmatic hemostasis. *Laboratory Medicine* 2008; 39: 225-30.

[44] Rivers EP, Coba V, Visbal A, Whitmill M, Amponsah D. Management of sepsis: early resuscitation. *Clin Chest Med* 2008; 29: 689-704.

[45] Muta T, Seki N, Takaki Y, Hashimoto R, Oda T, Iwanaga A, Tokunaga F, Iwaki D, Iwanaga S. Horseshoe crab factor G: a new heterodimeric serine protease zymogen sensitive to $(1\rightarrow3)$ß-D-glucan. *Advances in Experimental Medicine and Biology* 1996; 389: 79-85.

[46] Stief TW. Lipopolysaccharide reactivity of normal plasma. *Hemost Lab.* 2009; 2: 213-21.

[47] Stief T. An "APTT" for LMWH. *Hemost Lab.* 2013; 6: 367-72.

[48] Stief TW. Thrombin generation in presence of therapeutic antithrombin 3. *Hemost Lab.* 2012; 5: 91-102.

Chapter 3

ULTRAVIOLET A PHOTONS IN THE LATE PHASE OF OXIDATIVE BURST

ABSTRACT

Background: Blood neutrophils communicate by photons. They generate and perceive light of wavelengths in the range of about 300-400 nm (ultraviolet to blue). Is there a temporal relationship between oxidative burst and appearance of certain photons?

Material and Methods: 15 µl citrated blood were added to 225 µl HBSS in black polystyrene Brand®781608 F-wells. 0.26 mM luminol and 3.8 µg/ml zymosan A (ZyA) were added. Neighbor wells (NW) were filled with unstimulated blood (HBSS without luminol, without ZyA), or NW were empty. The blood ROS generation assay (BRGA) was performed and the reactive oxygen species (ROS) were measured by a photons-multiplying microtiter plate luminometer.

Results and Discussion: The longer the oxidative burst lasts the more photons are generated that do not pass to the neighbor well, as well NW = empty as NW = unstimulated blood. In unstimulated blood about half the photons are found compared to empty wells, the erythrocytes seem to absorb these photons. In strongly illuminated blood samples the photon receptors are impaired, subsequent triggering results in decreased blood ROS generation. Slightly illuminated blood samples are 2-4fold stimulated (primed). Physiological light upregulates light emission!

INTRODUCTION

Blood neutrophils are our main defense cells against fungi, bacteria, or micro-thrombi [1-7]. They generate strong oxidants of the type of hydrogen peroxide (H_2O_2), hydroxyl-radical ($\cdot OH$), and singlet oxygen ($^1O_2{}^*$). The latter oxidant is excited (*), i.e. it releases a photon upon returning to ground state oxygen. Normally, these photons are released out of oxygen bound in excited carbonyls ($R\text{-}C{=}O^*$), the wavelength (mainly 300-400 nm) of the resulting light depends on R and on the blood matrix. The neutrophils communicate by photons. They generate and perceive ultraviolet to blue light during their NADPH-oxidase assembly (oxidative burst). Is there a temporal relationship between oxidative burst and appearance of certain photons?

MATERIAL AND METHODS

15 µl venous blood supplemented with 10.6 mM sodium citrate in polypropylene monovettes (Sarstedt, Nümbrecht, Germany) of healthy donors that gave written informed consent were added in 8fold to Hanks' Balanced Salt Solution (HBSS; Sigma, Deisenhofen, Germany) in transparent polystyrene F-wells (Brand, Wertheim, Germany; article nr. 781602; final volume: 285 µl). 0.26 mM luminol sodium salt (in 15 µl 0.9% NaCl) (Sigma) and 3.8 µg/ml zymosan A (ZyA ; in 15 µl 0.9% NaCl) (Sigma) were added. Neighbor wells (NW) were filled in 8fold with 15 µl unstimulated blood (added to HBSS without luminol, without ZyA), or NW in 8fold were empty. The blood ROS generation assay (BRGA) was performed and the reactive oxygen species (ROS) were measured by a photons-multiplying microtiter plate luminometer (LUmo; Autobio-anthos, Krefeld, Germany) with 0.5s/well integration time. At 97 min (37°C) the ZyA concentration was increased from 3.8 to 7.2 µg/ml and unstimulated blood/HBSS were added as NW. After 30 min (37°C) unstimulated blood (either of illuminated wells or unilluminated wells) was supplemented with 0.26 mM luminol and 3.8 µg/ml ZyA, and the resulting luminescence was monitored.

RESULTS AND DISCUSSION

Figures 1-6 demonstrate the reaction kinetic of blood ROS generation. The longer the oxidative burst lasts the more photons are generated that cannot not pass to the neighbor well, as well NW = empty as NW = unstimulated blood. In unstimulated blood about half the photons are found compared to empty welly, the erythrocytes seem to absorb these photons.

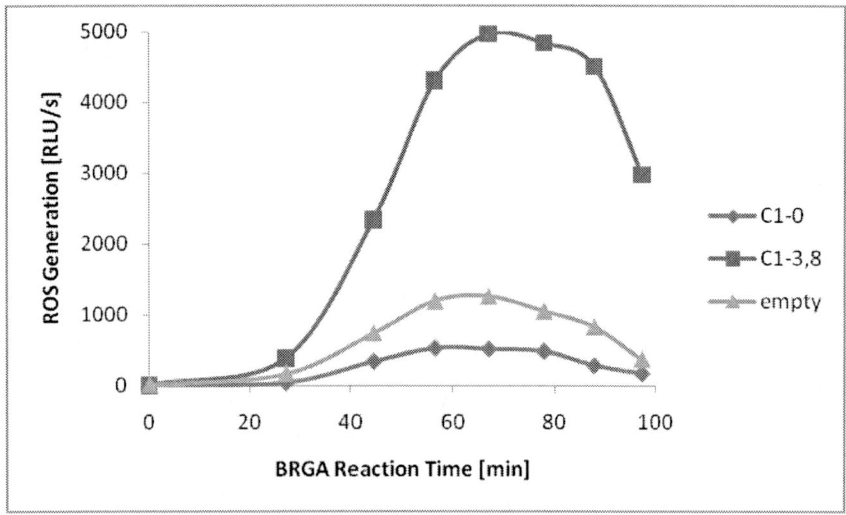

Figure 1. Reaction kinetic of BRGA. Citrated blood sample 1 (C1-3.8), triggered by 3.8 µg/ml ZyA (enhanced by 0.26 mM luminol), was measured in the BRGA. The neighbor wells (NW) contained unstimulated blood sample C1 (C1-0) or were empty.

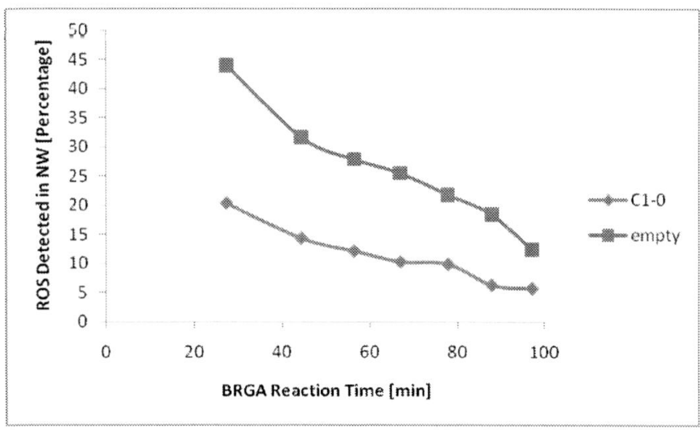

Figure 2. Percentage of ROS detected in neighbor wells. ROS were detected in NW of illuminating C1-3.8 wells (Figure 1) containing unstimulated blood (C1-0) or nothing.

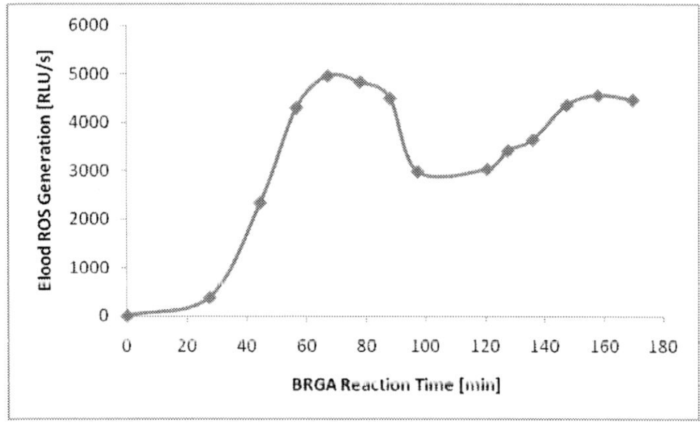

Figure 3. Reaction kinetic of BRGA upon twofold stimulation. Citrated blood was triggered by 3.8 µg/ml ZyA (enhanced by 0.26 mM luminol) and measured in the BRGA. At 97 min the ZyA concentration was increased to 7.2 µg/ml.

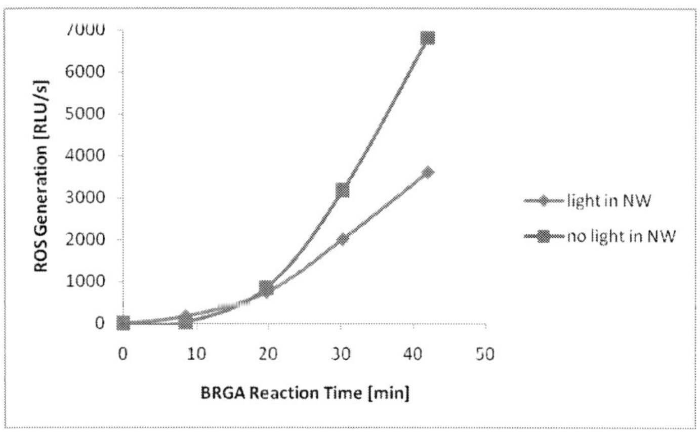

Figure 4. ROS generation in strongly illuminated wells. At 127 min reaction time point of figure 3 unstimulated blood samples, with light in NW and without light in NW, were stimulated by 3.8 µg/ml ZyA (0.26 mM luminol enhanced).

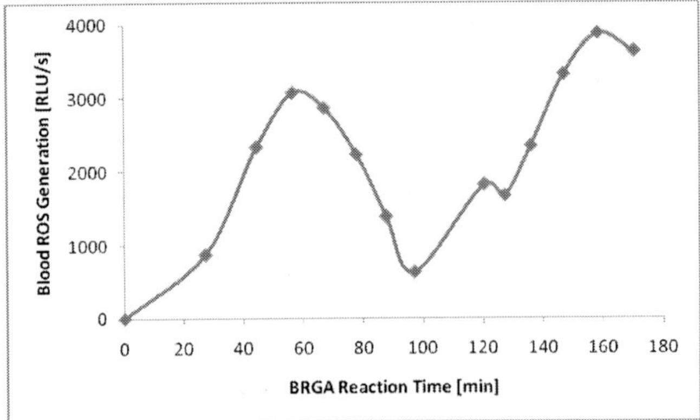

Figure 5. Reaction kinetic of BRGA upon twofold stimulation. Citrated blood was triggered in another micro plate by 3.8 µg/ml ZyA (enhanced by 0.26 mM luminol) and measured in the BRGA. At 97 min the ZyA concentration was increased to 7.2 µg/ml.

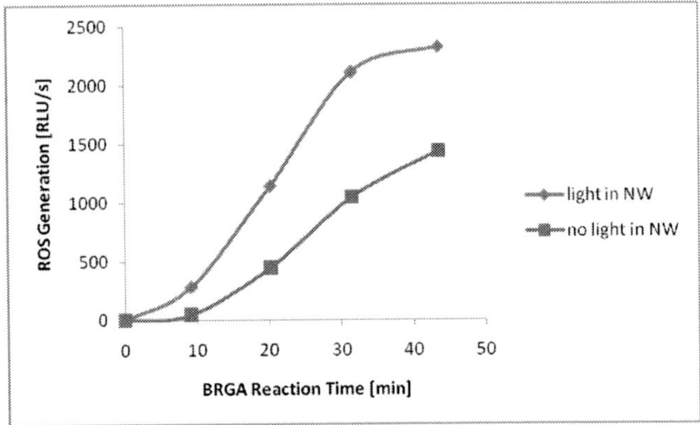

Figure 6. ROS generation in slightly illuminated wells. At 97 min reaction time point of figure 5 unstimulated blood C1 was added to the plate, at 127 min time point, with light in NW and without light in NW, the samples were stimulated by 3.8 µg/ml ZyA (0.26 mM luminol enhanced).

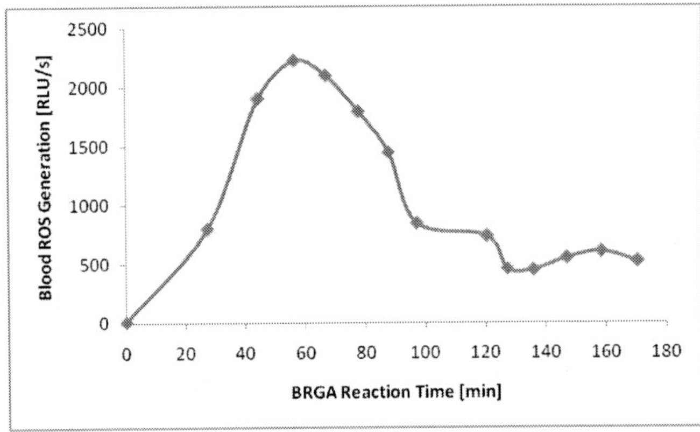

Figure 7. Reaction kinetic of BRGA upon twofold stimulation. Citrated blood C2 was triggered by 3.8 µg/ml ZyA (enhanced by 0.26 mM luminol) and measured in the BRGA. At 97 min the ZyA concentration was increased to 7.2 µg/ml.

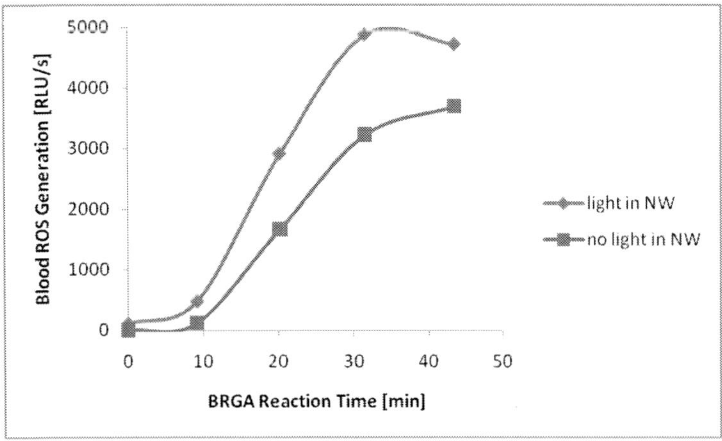

Figure 8. ROS generation in slightly illuminated wells. At 97 min reaction time point of figure 7 unstimulated blood C2 was added to the plate, at 127 min time point, with light in NW and without light in NW, the samples were stimulated by 3.8 µg/ml ZyA (0.26 mM luminol enhanced).

In strongly illuminated blood samples the opsin receptors for photons are impaired, subsequent triggering results in decreased blood ROS generation (Figures 4,5). Slightly illuminated blood samples are 2-4fold stimulated (primed), depending on the reaction time (Figures 5-8). Physiological light upregulates light emission!

REFERENCES

[1]   Stief T. *Photonic hemostasis. Physiology of Light Signals in the Neutrophil.* Nova Science Publishers, New York, 2013.

[2]   Stief T. Thrombin and Singlet Oxygen ($^1\Delta O_2^*$) *Main Factors of Hemostasis.* Nova Science Publishers, New York, 2013.

[3]   Freebern WJ, Bigwarfe TJ, Price KD, Haggerty HG. Methods: implementation of in vitro and ex vivo phagocytosis andrespiratory burst function assessments in safety testing. J Immunotoxicol. 2013; 10: 106-17.

[4]   Peluso I, Morabito G, Urban L, Ioannone F, Serafini M. Oxidative stress in atherosclerosis development: the central role of LDL and oxidative burst. Endocr Metab Immune Disord Drug Targets. 2012; 12: 351-60.

[5]   Lassmann H, van Horssen J, Mahad D. Progressive multiple sclerosis: pathology and pathogenesis. Nat Rev Neurol. 2012; 8: 647-56.

[6]   Maghzal GJ, Krause KH, Stocker R, Jaquet V. Detection of reactive oxygen species derived from the family of NOX NADPH oxidases. Free Radic Biol Med. 2012; 53: 1903-18.

[7]   Steevels TA, van Avondt K, Westerlaken GH, Stalpers F, Walk J, Bont L, Coffer PJ, Meyaard L. Signal inhibitory receptor on leukocytes-1 (SIRL-1) negatively regulates the oxidative burst in human phagocytes. Eur J Immunol. 2013; 43: 1297-308.

KALLIKREIN MODULATES SINGLET OXYGEN GENERATION

ABSTRACT

Background: Kallikrein is a very important starter enzyme of intrinsic hemostasis. Neutrophils react towards altered matrices with activation of their NADPH-oxidase and generation of reactive oxygen species (ROS). H_2O_2 is the original ROS that is converted via myeloperoxidase to the non-radicalic singlet oxygen ($^1\Delta O_2^*$). The regulation of $^1\Delta O_2^*$ generation by kallikrein is the purpose of this work.

Material and Methods: Freshest citrated blood or EDTA-blood (10 µl) were incubated with 120 µl Hanks' Balanced Salt Solution (HBSS without phenol red), 30 µl 0-100 ng/ml (final conc.) human plasma kallikrein in 5% human albumin, 10 µl 0.28 mM luminol (final conc.), 10 µl 0 or 2 µg/ml zymosan A (final conc.) in a black polystyrene plate (Brand®781608). The luminescence was measured by a photons-multiplying microtiter plate luminometer with an integration time of 0.5s per well.

Results and Discussion: The approx. IC50 was 0.4 ng/ml kallikrein at BRGA-26 (blood ROS generation assay with 26 min reaction time) for fresh citrated and EDTA-blood. At BRGA-146 appeared an approx. SC200 of 3 ng/ml kallikrein for fresh citrated blood. Thus, at short incubations kallikrein is an inhibitor and at long incubations kallikrein acts as stimulator of the neutrophil's singlet oxygen generation. This reaction is similar to that of the contact phase activator glucose. First the up-regulation of NADPH-oxidase assembly is hindered by kallikrein, possibly by cell hormone action of kallikrein/C1-inactivator complexes, later the down-regulation of the assembled NADPH-oxidase is inhibited.

Keywords: Kallikrein, reactive oxygen species, ROS, singlet oxygen, NADPH-oxidase, neutrophils

INTRODUCTION

Kallikrein is both an important tissue enzyme and one of the two starter enzymes of contact phase hemostasis (altered matrix coagulation) [1-7]. Neutrophils react towards altered matrices with activation of their NADPH-oxidase and generation of reactive oxygen species (ROS) [8]. H_2O_2 is the original ROS that is converted via myeloperoxidase to the non-radicalic singlet oxygen ($^1\Delta O_2^*$) [9-11]. The regulation of cellular $^1\Delta O_2^*$ generation [9-17] by kallikrein is the purpose of the present work.

MATERIAL AND METHODS

10 µl freshest (less than 0.5h old) individual normal venous blood of a healthy donor drawn from a left hand vein into polypropylene monovettes (4.5 ml blood added to 0.5 ml 106 mM sodium citrate, pH 7.4 or 2.6 ml blood supplemented with 1.6 mg/ml K_3-EDTA; Sarstedt, Nümbrecht, Germany) were incubated with 120 µl Hanks´ Balanced Salt Solution (HBSS without phenol red; Sigma, Deisenhofen, Germany), 30 µl 0-100 ng/ml (final conc.) human plasma kallikrein (Sigma; article nr. K2638-50UG; stem solution 50 µg/ml in 29% glycerol) in 5% human albumin (CSL Behring, Marburg, Germany), 10 µl 0.28 mM luminol (final conc.) (Sigma), 10 µl 0 or 2 µg/ml zymosan A (final conc.) (Sigma) in a black polystyrene plate (Brand, Wertheim, Germany; article nr. 781608). The luminescence was measured by a photons-multiplying microtiter plate luminometer (LUmo; Autobio-anthos, Krefeld, Germany) with an integration time of 0.5s per well [17,18]. The approximate 200% stimulatory or 50% inhibitory concentrations of kallikrein on blood ROS generation were determined (approx. SC200, approx. IC50).

RESULTS AND DISCUSSION

Figures 1-4 demonstrate the modulation of the generation of singlet oxygen by kallikrein. The approx. IC50 was 0.4 ng/ml kallikrein at BRGA-26 (blood ROS generation assay with 26 min reaction time) both for fresh citrated blood (Figure 1b) and fresh EDTA-blood (Figure 3b). At BRGA-105 there was no kallikrein-induced 50% reduction of singlet oxygen generation (Figure 3c). Instead, there appeared an approx. SC120 of 10 ng/ml kallikrein in citrated blood.

At BRGA-146 appeared an approx. SC200 of 3 ng/ml kallikrein for fresh citrated blood (Figure 2). For EDTA-blood the kallikrein-induced stimulation of singlet oxygen generation was less pronounced; 6 ng/ml kallikrein was the approx. SC125 at BRGA-146 (Figure 4). 0 µg/ml ZyA never resulted into the generation of singlet oxygen, i.e. kallikrein is not a trigger of NADPH-oxidase assembly, it may act as a primer under prolonged incubation conditions.

Thus, at short incubations kallikrein inhibits, and at long incubations kallikrein stimulates the neutrophil´s singlet oxygen generation. This reaction is similar to that of the contact phase activator glucose [19-23]. First the up-regulation of NADPH-oxidase assembly is hindered by kallikrein, possibly by cell hormone action of serine protease/serpin complexes (kallikrein/C1-inactivator) [24-28], later the down-regulation of the assembled NADPH-oxidase is inhibited [29].

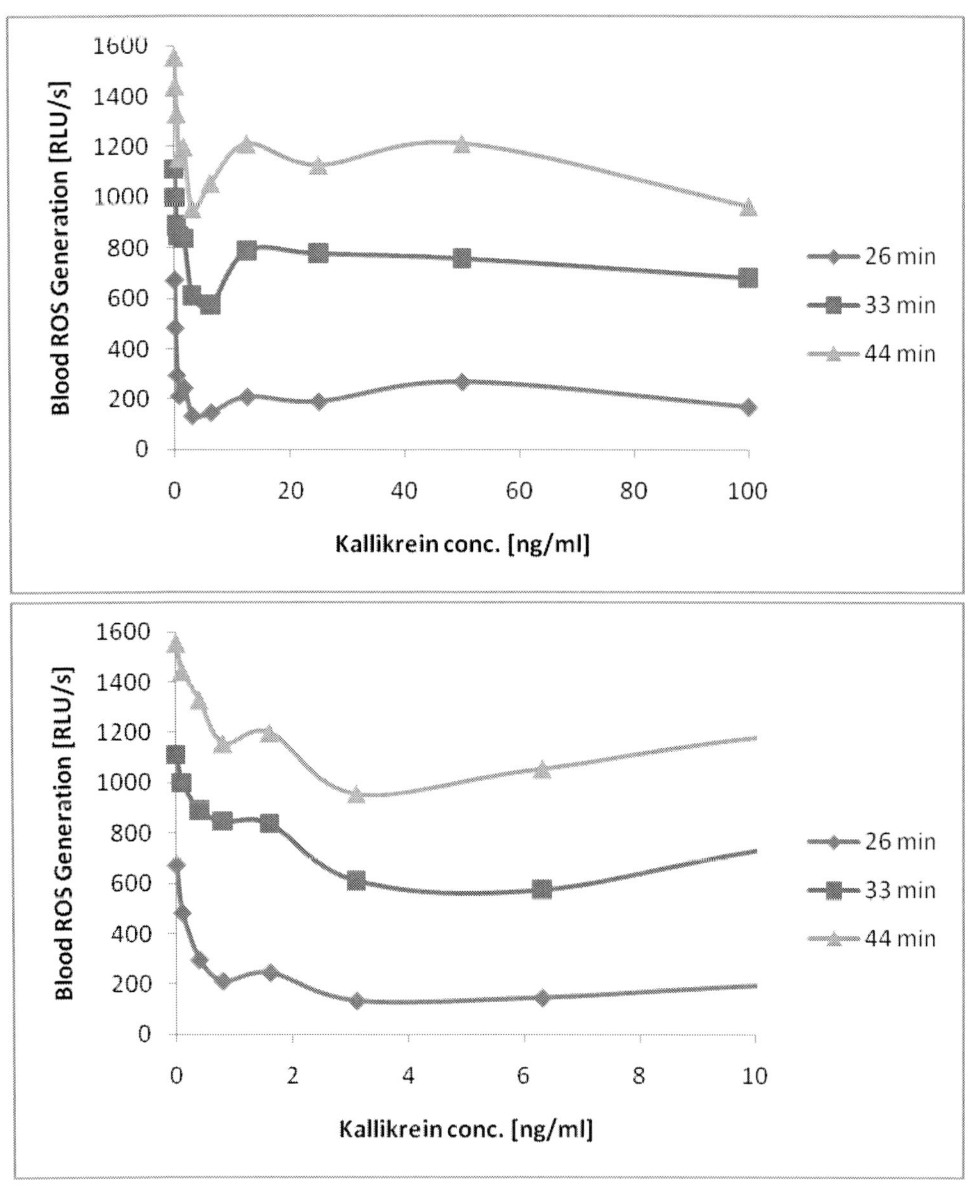

Figure 1. Singlet oxygen generation inhibition by kallikrein in initial neutrophil activation. Freshest citrated blood (10 µl) was incubated with 120 µl HBSS, 30 µl 0-100 ng/ml (final conc.) kallikrein in 5% human albumin, 10 µl 0.28 mM luminol (final conc.), 10 µl 2 µg/ml zymosan A (final conc.) in black polystyrene wells (Brand®781608). The luminescence (at 26, 33, 44 min) was measured by a microtiter plate luminometer. The ROS generation maximum (about 1900 RLU/s without kallikrein addition) occurred at about 1h (37°C). The approx. IC50 was 0.4 ng/ml kallikrein at BRGA-26.

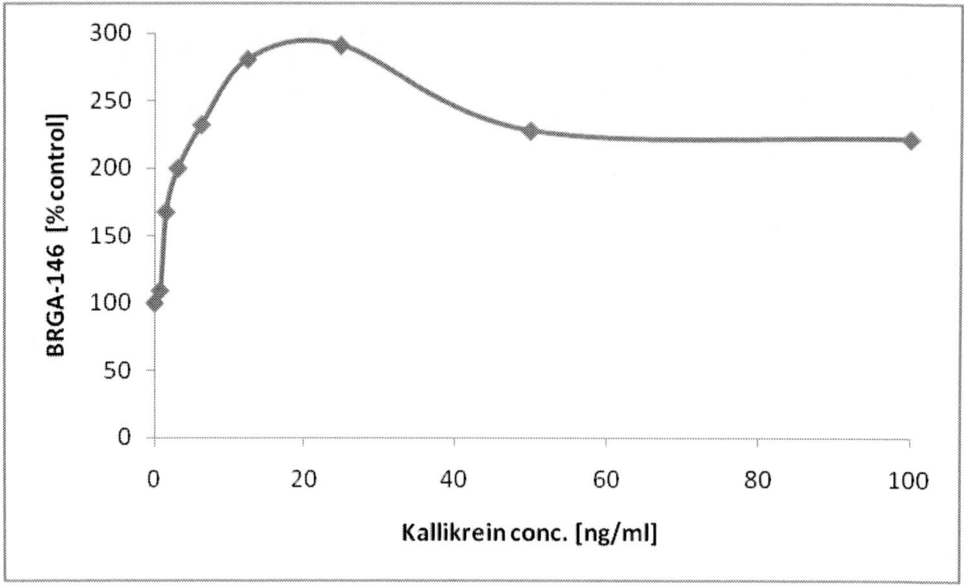

Figure 2. Singlet oxygen generation stimulation by kallikrein in late neutrophil activation. Freshest citrated blood (10 µl) was incubated in duplicate with 120 µl HBSS, 30 µl 0-100 ng/ml (final conc.) kallikrein, 10 µl 0.28 mM luminol (final conc.), 10 µl 2 µg/ml zymosan A in black polystyrene wells. The luminescence at 146 min (37°C; 100% = about 800 RLU/s) was measured by a photons-multiplying microtiter plate luminometer. The approx. SC200 was 3 ng/ml kallikrein at BRGA-146.

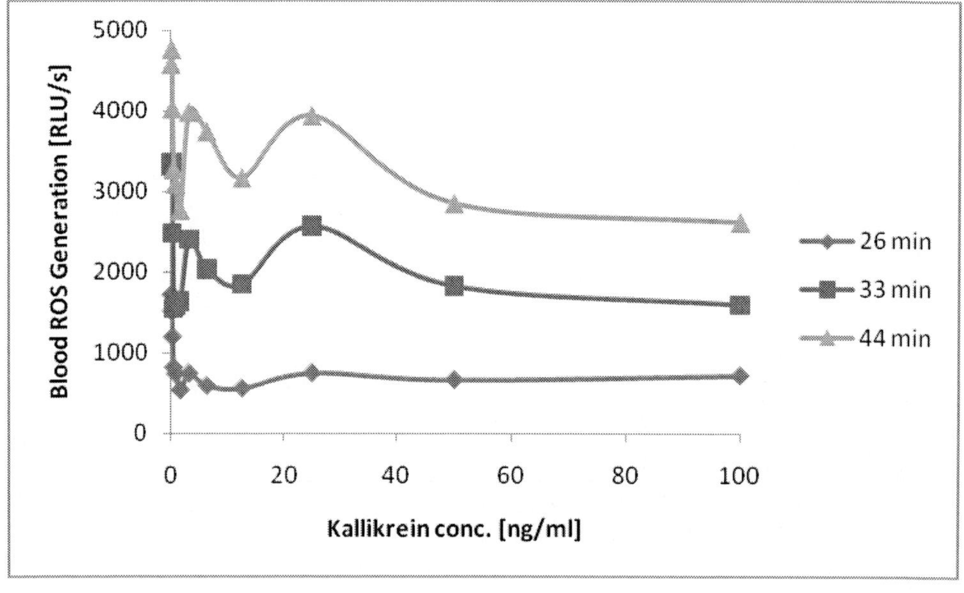

a

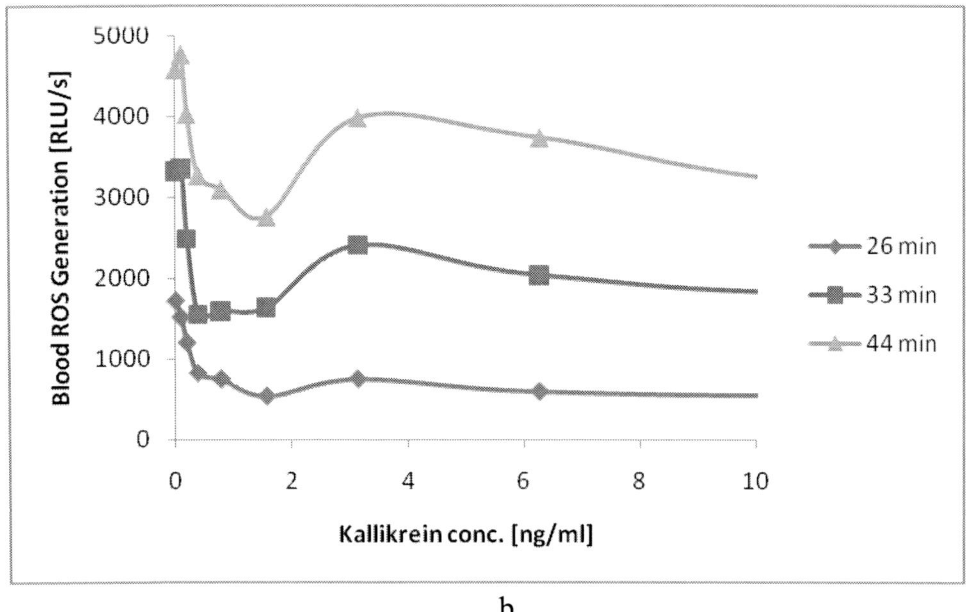

b

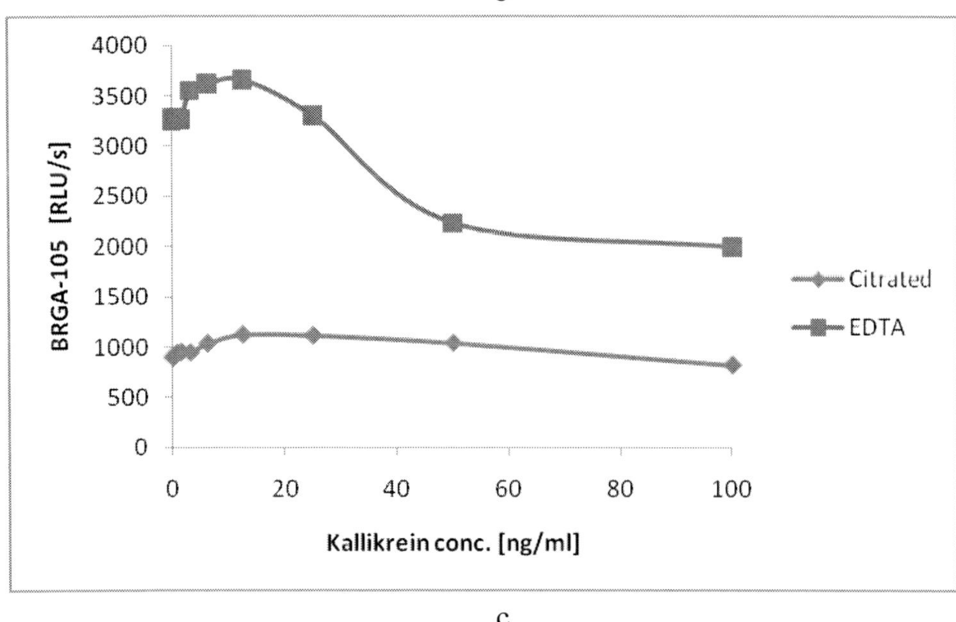

c

Figure 3. Singlet oxygen generation inhibition by kallikrein in initial neutrophil activation. Freshest EDTA-blood (10 µl) was incubated with 120 µl HBSS, 30 µl 0-100 ng/ml (final conc.) kallikrein, 10 µl 0.28 mM luminol (final conc.), 10 µl 2 µg/ml zymosan A (final conc.) in black polystyrene wells (Brand®781608). The luminescence (at 26, 33, 44 min) was measured by a photons-multiplying microtiter plate luminometer. The ROS generation maximum (about 6000 RLU/s without kallikrein addition) occurred at about 1h (37°C). The approx. IC50 was 0.4 ng/ml kallikrein at BRGA-26 (Figure 3a, 3b). Figure 3c shows absence of IC50 in BRGA-105 (blood ROS generation assay with 105 min reaction time), especially in citrated blood.

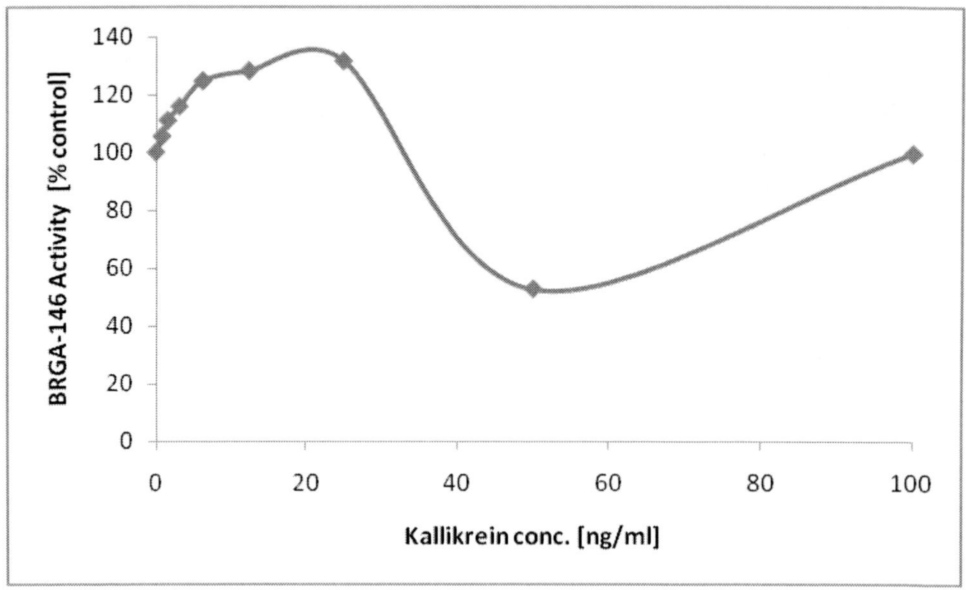

Figure 4. Singlet oxygen generation stimulation by kallikrein in late neutrophil activation. Freshest citrated blood (10 µl) was incubated in duplicate with 120 µl HBSS, 30 µl 0-100 ng/ml (final conc.) kallikrein, 10 µl 0.28 mM luminol, 10 µl 2 µg/ml zymosan A in black polystyrene wells (Brand®781608). The luminescence at 146 min (37°C; 100% = about 2500 RLU/s) was measured by a photons-multiplying microtiter plate luminometer. The approximate 125% stimulatory conc. (SC125) was 6 ng/ml kallikrein at BRGA-146.

REFERENCES

[1]   Stief TW. Kallikrein activates prothrombin. *Clin Appl Thrombosis/Hemostasis* 2008: 14: 97-8.

[2]   Stief TW. Kallikrein triggers thrombin generation. *Hemostasis Laboratory* 2009; 2: 45-56.

[3]   Stief TW. Zn^{2+}, hexane, or glucose activate factor 12 and/or prekallikrein in two purified systems. *Hemostasis Laboratory* 2011; 4: 409-26.

[4]   Stief TW. Kallikrein activates factor 10. *Hemostasis Laboratory* 2012; 5: 211-8.

[5]   Stief TW, Klingmüller V. Diagnostic ultrasound activates pure prekallikrein. *Blood Coagulation and Fibrinolysis* 2012; 23: 781-3.

[6]   Stief T. Hemostasis activation in inflammation: kallikrein, thrombin, $^{1}\Delta O_{2}^{*}$/hv. In: *Thrombin and Singlet Oxygen ($^{1}\Delta O_{2}^{*}$) Main Factors of Hemostasis*. Stief T, ed.; Nova Science Publishers, New York, 2013.

[7]   Stief TW. Coumarins trigger altered matrix (AM) coagulation activation. *Hemost Lab.* 2013; 6: 121-8.

[8]   Wachtfogel YT, Kettner C, Hack CE, Nuijens JH, Reilly TM, Knabb RM, Kucich U, Niewiarowski S, Edmunds LH Jr, Colman RW. Thrombin and human plasma kallikrein inhibition during simulated extracorporeal circulation block platelet and neutrophil activation. *Thromb Haemost.* 1998; 80: 686-91.

[9]   Stief TW. Neutrophil granulocytes in hemostasis. *Hemostasis Laboratory* 2008; 1: 269-89.

[10] Nathan C. Neutrophils and immunity: challenges and opportunities. *Nat Rev Immunol.* 2006; 6: 173-82.

[11] Stief TW. The blood fibrinolysis / deep-sea analogy: a hypothesis on the cell signals singlet oxygen/photons as natural antithrombotics. *Thromb Res* 2000; 99: 1-20.

[12] Stief TW, Fareed J. The antithrombotic factor singlet oxygen/light ($^1O_2$/hv). *Clin Appl Thrombosis/Hemostasis* 2000; 6: 22-30.

[13] Stief TW, Fu K, Yang LH, Ramaswamy A, Fareed J. Singlet oxygen ($^1O_2$) induces selective thrombolysis in vivo by massive granulocyte infiltration into the thrombus. 43. GTH Congress, Mannheim, 24.-27.2.1999, *Ann. Hematol.* 1999; 78: A32 (FV112).

[14] Stief TW. The physiology and pharmacology of singlet oxygen. *Med Hypoth* 2003; 60: 567-572.

[15] Stief TW. Regulation of hemostasis by singlet oxygen ($^1\Delta O_2$). *Curr Vasc Pharmacol* 2004; 2: 357-362.

[16] Stief T. Micro-thrombi stimulate blood ROS generation. *Hemostasis Laboratory* 2013; 6: 315-25.

[17] Stief T. The routine blood ROS generation assay (BRGA) triggered by typical septic concentrations of zymosan A. *Hemostasis Laboratory* 2013; 6: 89-98.

[18] Stief T. Glucose initially inhibits and later stimulates blood ROS generation. *Journal of Diabetes Mellitus* 2013, 3: 15-21.

[19] Stief TW. Glucose triggers thrombin generation. *Hemostasis Laboratory* 2010; 3: 93-103.

[20] Stief TW. Zn^{2+}, hexane, or glucose activate factor 12 and/or prekallikrein in two purified systems. *Hemostasis Laboratory* 2011; 4: 409-26.

[21] Stief TW. Glucose activates the early phase of intrinsic coagulation. *Hemostasis Laboratory* 2012; 5: 67-81.

[22] Stief TW, Mohrez M. Glucose activates human intrinsic coagulation *in vivo*. *Hemostasis Laboratory* 2012; 5: 83-9.

[23] Stief TW, Mohrez M. Glucose provokes pathologic plasmatic thrombin generation. In: *Thrombin: function and pathophysiology*. Stief T, ed.; NOVA science publishers; New York; 2012; pp.23-35.

[24] Bar-Shavit R, Wilner GD. Biologic activities of nonenzymatic thrombin: elucidation of a macrophage interactive domain. *Semin Thromb Hemost.* 1986; 12: 244-9.

[25] Herbert JM, Dupuy E, Laplace MC, Zini JM, Bar Shavit R, Tobelem G. Thrombin induces endothelial cell growth via both a proteolytic and a non-proteolytic pathway. *Biochem J.* 1994; 303: 227-31.

[26] Malek R, Aulak KS, Davis AE 3rd. The catabolism of intact, reactive centre-cleaved and proteinase-complexed C1 inhibitor in the guinea pig. *Clin Exp Immunol.* 1996; 105: 191-7.

[27] Davis AE 3rd. Structure and function of C1 inhibitor. *Behring Inst Mitt.* 1989; 84:142-50.

[28] Stief TW, Schorlemmer HU, Beck-Speier I, Doss MO. PAI-2 inhibits the chemiluminescence of phagocytes and suppresses autoimmunity. *Fibrinolysis and Proteolysis* 1999; 13: 245-51.

[29] Stanton RC. Glucose-6-phosphate dehydrogenase, NADPH, and cell survival. *IUBMB Life.* 2012; 64: 362-9.

Chapter 5

METAMIZOLE IS A STRONG TRIGGER OF INTRINSIC COAGULATION

ABSTRACT

Background: Metamizole is an important analgetic and antipyretic drug. It contains a sulfonate group and a benzene ring, i.e. the molecule has strongly negatively charged and strongly lipophilic parts.

Negatively charged or lipophilic molecules trigger intrinsic hemostasis, folding factor 12 or pre-kallikrein into F12a or kallikrein. The best assay to determine even spurious activations of contact phase coagulation is the recalcified coagulation activity assay (RECA).

Material and Methods: 40 µl platelet poor citrated plasma of 5 healthy donors were supplemented with 0-119 mg/l metamizole by repetitive 1+1 dilution on the polystyrene microtiter plate (Brand®781600). The RECA was performed with 12 or 20 min (37°C) coagulation reaction time. The approximate 200% stimulatory concentration (approx. SC200) was determined for each individual plasma.

Results and Discussion: Metamizole triggered intrinsic coagulation with approx. SC200 values of 4±2 mg/l.

This is a very small concentration, considering that plasma concentrations > 100 mg/l are therapeutically possible. This means that metamizole must be considered a pro-thrombotic risk factor in patients with further thrombosis risks. Then the individual patient who needs metamizole could benefit from a LMWH-prophylaxis.

INTRODUCTION

The analgetic and antipyretic drug metamizole [1] contains a sulfonate group and a benzene ring, i.e. metamizole has strongly negatively charged and strongly lipophilic parts [2]. Negatively charged or lipophilic molecules trigger intrinsic hemostasis, they fold factor 12 into F12a or pre-kallikrein into kallikrein [3-6]. The best assay to determine even spurious activations of contact phase coagulation is the recalcified coagulation activity assay (RECA) [7].

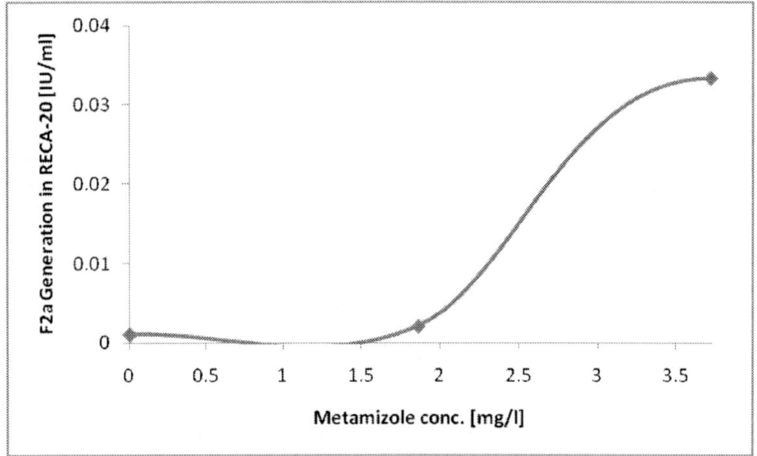

Figure 1. Chemical structure of metamizole. Sodium [(2,3-dihydro-1,5-dimethyl-3-oxo-2-phenyl-1*H*-pyrazol-4-yl)methylamino] methanesulfonate [2].

MATERIAL AND METHODS

40 µl platelet poor citrated plasma of 5 healthy donors that gave written informed consent were supplemented with 0-119 mg/l metamizole (ratiopharm, Ulm, Germany; ampoules of 1g) by repetitive 1+1 dilution on the polystyrene microtiter plate (Brand, Wertheim, Germany; article nr. 781600).

The RECA was started with 4 µl $CaCl_2$. After 0, 12, or 20 min (37°C) coagulation reaction time (CRT) the RECA was stopped by addition of 80 µl 2.5 M arginine, pH 8.6, 0.16% Triton X 100®. After 3 min 20 µl chromogenic thrombin substrate 1 mM HD-CHG-Ala-Arg-pNA in 1.25 M arginine, pH 8.7, were added and the increase in absorbance with time $\Delta A/t$ was determined at 405 nm by microtiter plate photometer with a 1 mA resolution (PHOmo; Autobio-anthos, Krefeld, Germany).

The approximate 200% stimulatory concentration (approx. SC200) was determined for each individual plasma. Considered were only the thrombin activities in the ascending part of the CRT vs. thrombin activity generation curve.

Figure 2. SC200 determination in plasma 1. 40 µl citrated plasma 1, supplemented with 0-119 mg/l metamizole, were recalcified and the thrombin generation was stopped by the arginine-reagent as described under Methods. The approx. SC200 was 2 mg/l.

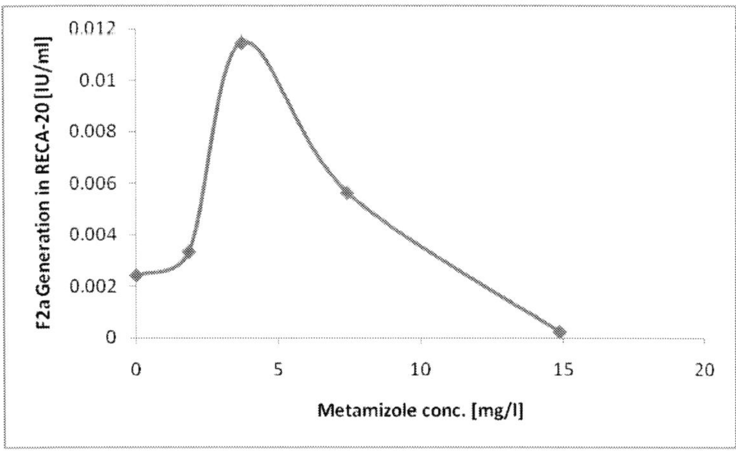

Figure 3. SC200 determination in plasma 2. 40 µl citrated plasma 2, supplemented with 0-119 mg/l metamizole, were recalcified and the thrombin generation was stopped by the arginine-reagent as described under Methods. The approx. SC200 was 3 mg/l.

RESULTS AND DISCUSSION

Figures 2-6 demonstrate that the approx. SC200 of metamizole on blood ROS generation is only 4±2 mg/l.

This is in the range of the approx. IC50 [8] of cytochrome P450 metabolized [9] metamizole on blood ROS generation. This is a very small concentration, considering that plasma concentrations > 100 mg/l are therapeutically possible and means that metamizole must be considered a pro-thrombotic risk factor in patients with further thrombosis risks. Then the individual patient who needs metamizole could benefit from a LMWH-prophylaxis [10].

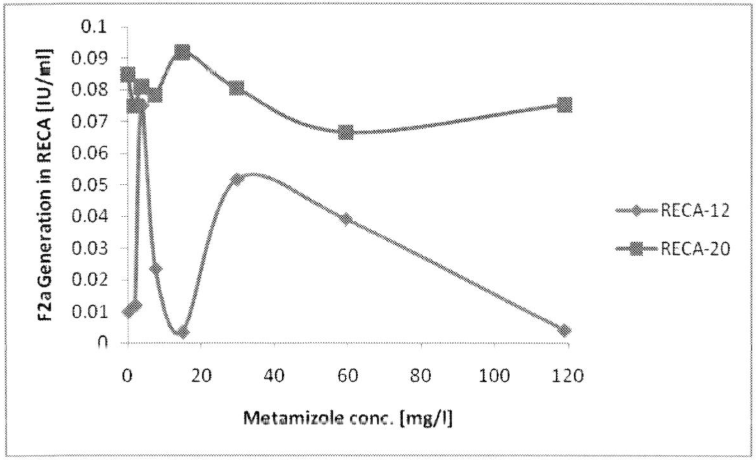

Figure 4. SC200 determination in plasma 3. 40 µl citrated plasma 3, supplemented with 0-119 mg/l metamizole, were recalcified and the thrombin generation was stopped by the arginine-reagent as described under Methods. The approx. SC200 was 3 mg/l.

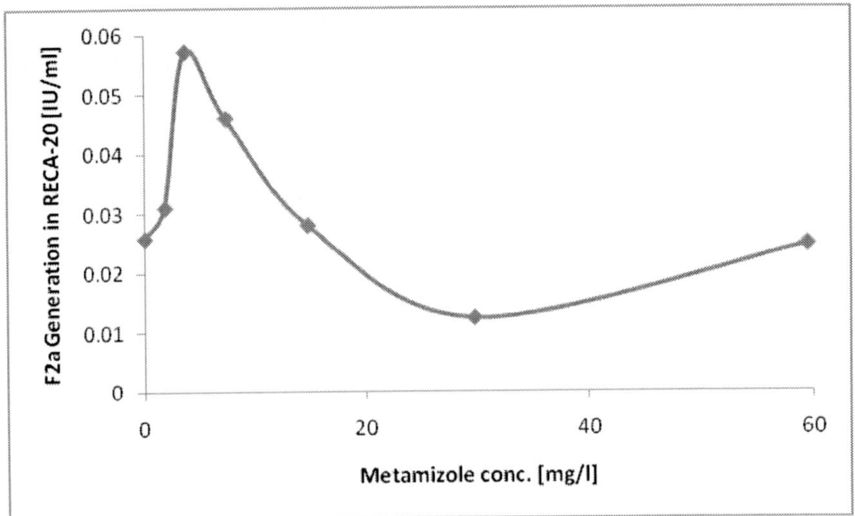

Figure 5. SC200 determination in plasma 4. 40 µl citrated plasma 4, supplemented with 0-119 mg/l metamizole, were recalcified and the thrombin generation was stopped by the arginine-reagent as described under Methods. The approx. SC200 was 4 mg/l.

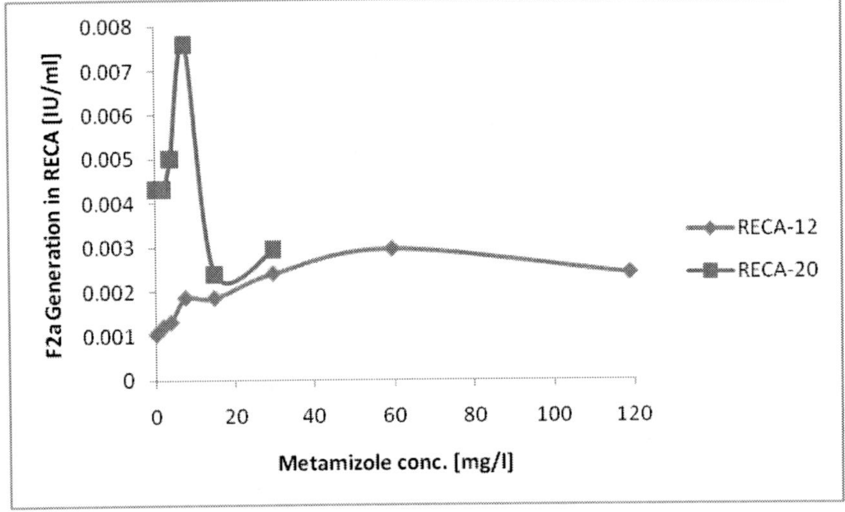

Figure 6. SC200 determination in plasma 5. 40 µl citrated plasma 5, supplemented with 0-119 mg/l metamizole, were recalcified and the thrombin generation was stopped by the arginine-reagent as described under Methods. The approx. SC200 was 8 mg/l.

REFERENCES

[1]   Pereira GL, Tavares NU, Mengue SS, Pizzol Tda S. Therapeutic procedures and use of alternating antipyretic drugs for fever management in children. *J Pediatr.* 2013; 89: 25-32.

[2]   www.wikipedia.org

[3] Stief TW. Factor 12 activation in two purified systems. *Hemostasis Laboratory* 2011; 4: 385-408.

[4] Stief TW. Zn^{2+}, hexane, or glucose activate factor 12 and/or prekallikrein in two purified systems. *Hemostasis Laboratory* 2011; 4: 409-26.

[5] Stief TW. Zn^{2+}, hexane, valproate, or glucose in two purified systems of F12-PK-HMWK. *Hemostasis Laboratory* 2012; 5: 35-50.

[6] Stief TW, Mohrez M. HMWK increases of decreases F12a generation dependent on the contact trigger concentration in two purified systems. *Hemostasis Laboratory* 2012; 5: 51-65.

[7] Stief TW. Pathological thrombin generation by the synthetic inhibitor argatroban. *Hemostasis Laboratory* 2009; 2: 83-104.

[8] Stief T. Pre-incubated metamizole suppresses blood ROS about tenfold stronger than metamizole. *Hemost Lab.* 2013; 7 (issue 2)

[9] Imaoka S, Inoue K, Funae Y. Aminopyrine metabolism by multiple forms of cytochrome P-450 from rat liver microsomes: simultaneous quantitation of four aminopyrine metabolites by high-performance liquid chromatography. *Arch Biochem Biophys.* 1988; 265: 159-70.

[10] Stief TW. LMWH - action-monitoring for all patients. *Acta Paediatr.* 2012; 101: e314.

Chapter 6

VITAMIN B$_1$ (THIAMINE) TRIGGERS AM-COAGULATION

ABSTRACT

Background: Thiamine is the hydrophilic vitamin B$_1$ that consists of amino-pyrimidine linked to thiazole. The normal blood concentration of vitamin B$_1$ is 20-100 µg/l, most of the vitamin is located inside the erythrocyte. Vitamin B$_1$ can clinically be given as 100 mg injections. Therefore, this work quantified the changed thrombin generation by alteration of matrix thru an increase of vitamin B$_1$ concentration.

Material and Methods: 40 µl platelet poor citrated plasma of 6 healthy donors were supplemented with 0-30.5 mg/l thiamine by repetitive 1+1 dilution on the polystyrene microtiter plate (Brand®781600). The RECA was performed with 10 or 20 min (37°C) coagulation reaction time. The approximate 200% stimulatory concentration (approx. SC200) was determined for each individual plasma.

Results and Discussion: Thiamine triggered AM-coagulation with approx. SC200 values of 2.0±1.6 mg/l. This is about 10fold less than the blood plasma concentrations reached directly after injection of 100 mg thiamine. This means that vitamin B$_1$ should be considered a pro-thrombotic risk factor in patients with further thrombosis risks. Then the individual patient who needs thiamine could benefit from oral administration or a combined vitamin B$_1$/LMWH treatment.

INTRODUCTION

Thiamine is a hydrophilic vitamin (vitamin B$_1$). It consists of an amino-pyrimidine ring linked to a thiazole ring (Figure 1) [1]. Especially the thiazole ring is important for the transfer of 2 carbon units. Thiamine diphosphate (TPP) is coenzyme in oxidative decarboxylation of 2-oxoacids (α-keto acids) and conjugation with coenzyme A (dehydrogenation) [2]. Typical enzymes are pyruvate dehydrogenase, 2-oxoglutarate dehydrogenase, branched-chain α-keto acid dehydrogenase, or transketolase (connecting the NADPH- generating pentose phosphate pathway to glycolysis). TPP is also used in the biosynthesis of the neurotransmitter acetylcholine and gamma-aminobutyric acid (GABA).

Approximately 90% of total thiamine in blood (about 20-100 µg/l) is inside the erythrocytes. Free thiamine and thiamine monophosphate can cross cell membranes. But cellular uptake of thiamine is enhanced via active transport, e.g. when passing the blood brain barrier. Neurons need much more thiamine for normal function than other cells. Thiamine deficiency poly-neuropathy with clinical signs of optic neuropathy, Wernicke-Korsakoff´s syndrome, peripheral polyneuritis, beriberi ("burning feet-syndrome"), sleepiness, or cardiovascular dysfunction. Therefore, thiamine was formerly named anti-neuritic vitamin (aneurin) [1,3].

Upon TPP-binding to messenger RNA (mRNA) the TPP metabolism or transfer is down-regulated (signaling that there is enough TPP inside the cell, more of it could be dangerous !). mRNA with bound TPP cannot be translated into proteins, the TPP induced mRNAstop [4].

The daily requirement of thiamine is about 1 mg, about 25-30 mg of thiamine are stored in skeletal muscle, heart, brain, liver, and kidneys. Thiamine, thiamine acetic acid, 2-methyl-4-amino-5-pyrimidine carboxylic acid, 4-methyl-thiazole-5-acetic acid, are renally eliminated.

Thiamine is clinically available in 100 mg amounts per ampoule. Since thiamine contains 3 delta-negatively charged nitrogens and one delta-negatively charged oxygen linked to lipophilic -CH_2-CH_2 thiamine could trigger the contact phase of human coagulation (= intrinsic coagulation = altered matrix (AM) coagulation); negatively charged or lipophilic molecules fold factor 12 into F12a or pre-kallikrein into kallikrein [5-8]. The best assay to determine even spurious activations of AM-coagulation is the recalcified coagulation activity assay (RECA) [9].

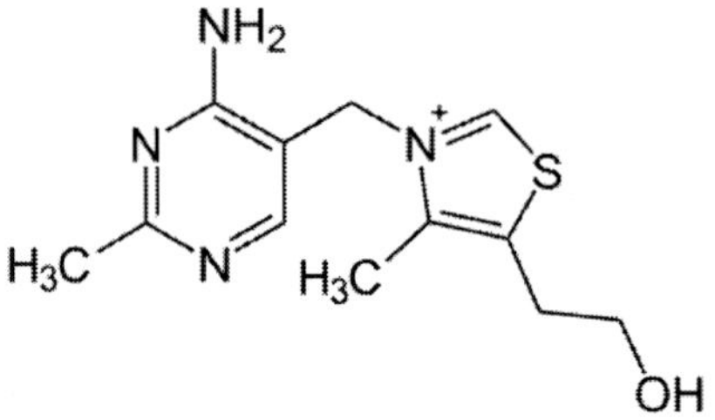

Figure 1. Chemical structure of vitamin B_1 (thiamine) [1]. Pyrimidine linked via methylene to thiazole ($C_{12}H_{18}ClN_4OS$; thiamin · HCl; molecular weight 332.3). Ester reaction of thiamine´s hydroxyl group with activated pyro-phosphoric acid (in adenosine triphosphate) transforms thiamine into the active metabolite thiamine diphosphate (TPP).

MATERIAL AND METHODS

40 µl -30°C frozen/23°C thawed platelet poor citrated plasma of 6 healthy donors that gave written informed consent were supplemented with 0-30.5 mg/l thiamin (ratiopharm, Ulm, Germany; 2 ml ampoules of 50 mg/ml) by repetitive 1+1 dilution on the polystyrene microtiter plate (Brand, Wertheim, Germany; article nr. 781600). The RECA was started with 4 µl 250 mM $CaCl_2$ (Sigma, Deisenhofen, Germany) After 0, 10, or 20 min (37°C) coagulation reaction time (CRT) the RECA was stopped by addition of 80 µl 2.5 M arginine, pH 8.6, 0.16% Triton X 100® (Sigma). After 3 min 20 µl chromogenic thrombin substrate 1 mM HD-CHG-Ala-Arg-pNA (Pentapharm, Basel, Switzerland) in 1.25 M arginine, pH 8.7, were added and the increase in absorbance with time ΔA/t was determined at 405 nm by a microtiter plate photometer with a 1 mA resolution (PHOmo; Autobio-anthos, Krefeld,

Germany). The approximate 200% stimulatory concentration (approx. SC200) or the approximate 50% inhibitory concentration (approx. IC50) was determined for each individual plasma at the most sensible thrombin generation curve. Considered were only the thrombin activities in the ascending part of the CRT vs. thrombin activity generation curve.

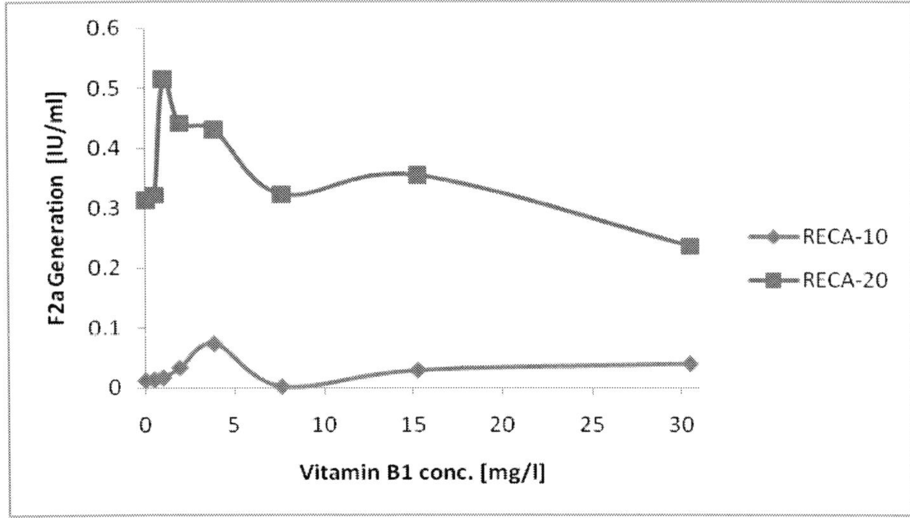

Figure 2. SC200 determination in plasma 1. 40 µl citrated plasma 1, supplemented with 0-30.5 mg/l vitamin B$_1$, were recalcified and the thrombin generation was stopped after 10 min or 20 min by arginine as described under Methods. The approx. SC200 was 1.5 mg/l.

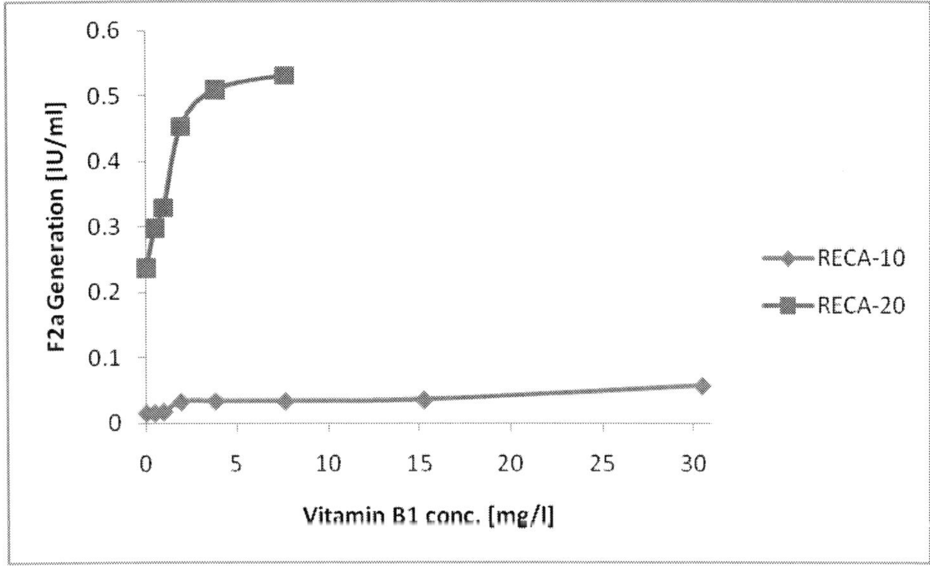

Figure 3. SC200 determination in plasma 2. 40 µl citrated plasma 2, supplemented with 0-30.5 mg/l vitamin B$_1$, were recalcified and the thrombin generation was stopped after 10 min or 20 min by arginine as described under Methods. The approx. SC200 was 2 mg/l.

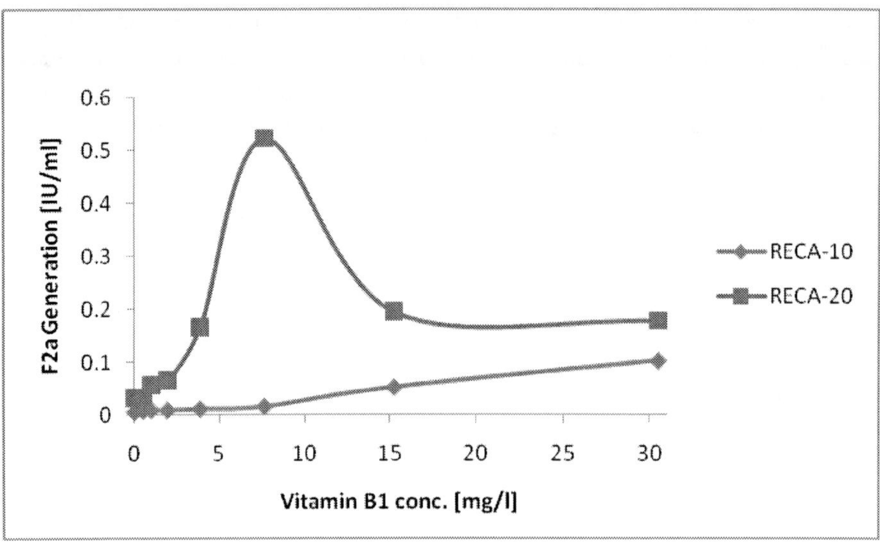

Figure 4. SC200 determination in plasma 3. 40 μl citrated plasma 3, supplemented with 0-30.5 mg/l vitamin B₁, were recalcified and the thrombin generation was stopped after 10 min or 20 min by arginine as described under Methods. The approx. SC200 was 2 mg/l.

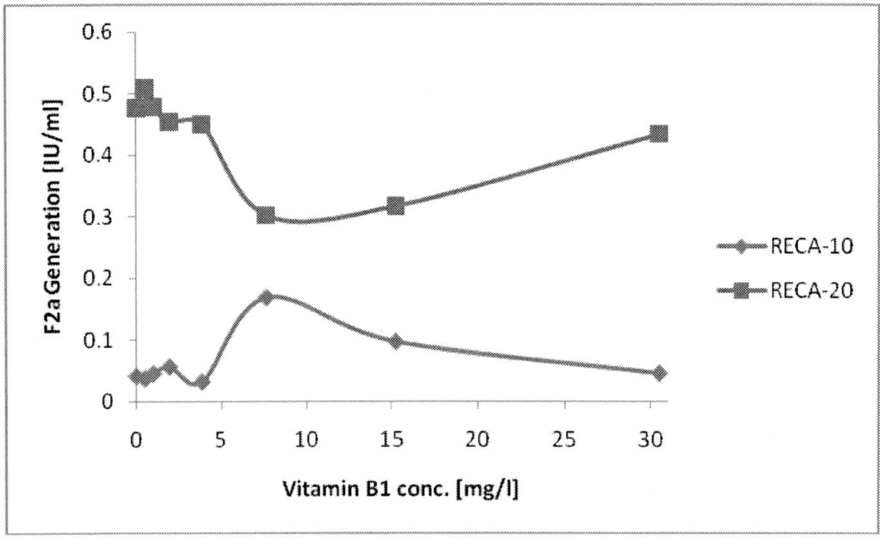

Figure 5. SC200 determination in plasma 4. 40 μl citrated plasma 4, supplemented with 0-30.5 mg/l vitamin B₁, were recalcified and the thrombin generation was stopped after 10 min or 20 min by arginine as described under Methods. The approx. SC200 was 5 mg/l.

RESULTS AND DISCUSSION

Figures 2-7 demonstrate that the approx. SC200 on AM-coagulation was 2.0±1.6 mg/l (MV±1SD) vitamin B₁ (range: 0.4-5 mg/l). An injection of 100 mg vitamin B₁ would instantaneously result in a plasma concentration of about 30 mg/l vitamin B₁. This is more than 10fold the mean value of the approx. SC200 on AM-coagulation. Therefore, vitamin B₁ should not be administered intravenously, oral vitamin B_1 has a much slower increase of plasma concentration. If there are really clinical situations where vitamin B_1 injection is necessary, thiamine should be accompanied by low-molecular-weight-heparin (LMWH) thrombosis prophylaxis [10].

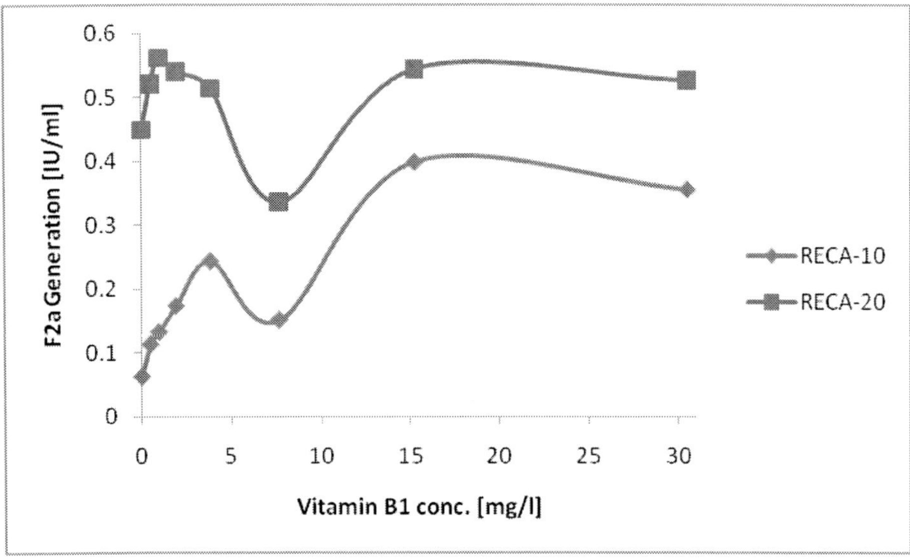

Figure 6. SC200 determination in plasma 5. 40 µl citrated plasma 5, supplemented with 0-30.5 mg/l vitamin B_1, were recalcified and the thrombin generation was stopped after 10 min or 20 min by arginine as described under Methods. The approx. SC200 was 1 mg/l.

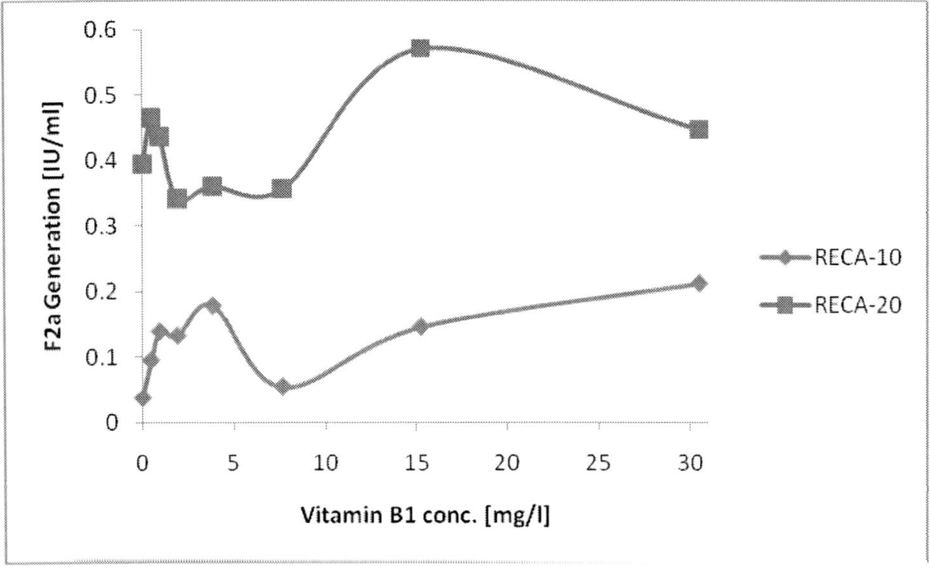

Figure 7. SC200 determination in plasma 6. 40 µl citrated plasma 6, supplemented with 0-30.5 mg/l vitamin B_1, were recalcified and the thrombin generation was stopped after 10 min or 20 min by arginine as described under Methods. The approx. SC200 was 0.4 mg/l.

REFERENCES

[1] www.wikipedia.org

[2] Li T, Huo L, Pulley C, Liu A. Decarboxylation mechanisms in biological system. Bioorg Chem. 2012; 43: 2-14.

[3] Osiezagha K, Ali S, Freeman C, Barker NC, Jabeen S, Maitra S, Olagbemiro Y, Richie W, Bailey RK. Thiamine deficiency and delirium. Innov Clin Neurosci. 2013; 10: 26-32.

[4] Serganov, A; Polonskaia A, Phan AT, Breaker RR, Patel DJ. Structural basis for gene regulation by a thiamine pyrophosphate-sensing riboswitch. *Nature.* 2006; 441: 1167–71.

[5] Stief TW. Factor 12 activation in two purified systems. *Hemostasis Laboratory* 2011; 4: 385-408.

[6] Stief TW. Zn^{2+}, hexane, or glucose activate factor 12 and/or prekallikrein in two purified systems. *Hemostasis Laboratory* 2011; 4: 409-26.

[7] Stief TW. Zn^{2+}, hexane, valproate, or glucose in two purified systems of F12-PK-HMWK. *Hemostasis Laboratory* 2012; 5: 35-50.

[8] Stief TW, Mohrez M. HMWK increases of decreases F12a generation dependent on the contact trigger concentration in two purified systems. *Hemostasis Laboratory* 2012; 5: 51-65.

[9] Stief TW. Pathological thrombin generation by the synthetic inhibitor argatroban. *Hemostasis Laboratory* 2009; 2: 83-104.

[10] Stief TW. LMWH - action-monitoring for all patients. Acta Paediatr. 2012; 101: e314.

Chapter 7

VITAMIN B$_2$ (RIBOFLAVIN) TRIGGERS AM-COAGULATION

ABSTRACT

Background: Riboflavine (vitamin B$_2$) is the central component of the cofactors flavin adenine dinucleotide (FAD) and flavin mononucleotide (FMN). Since riboflavine contains 2 delta-negatively charged oxygens in keto-position, 4 delta-negatively charged oxygens in hydroxyl-groups, and a lipophilic benzene ring with 2 methyl-groups, riboflavin could trigger altered matrix (AM) - coagulation; negatively charged or lipophilic molecules fold factor 12 into F12a or pre-kallikrein into kallikrein. The best assay to determine even spurious activations of AM-coagulation is the recalcified coagulation activity assay (RECA).

Material and Methods: 40 µl platelet poor citrated plasma of 6 healthy donors were supplemented with 0-3.05 mg/l vitamin B$_2$ by repetitive 1+1 dilution on the polystyrene microtiter plate (Brand®781600). The RECA was performed with 10 or 20 min (37°C) coagulation reaction time. The approx. SC200 or approx. IC50 was determined for each individual plasma.

Results and Discussion: The approx. SC200 on AM-coagulation was 0.1±0.04 mg/l (MV±1SD) vitamin B2 (range: 0.04-1.5 mg/l) in 5 of 6 samples. The 6th sample had an approx. IC50 of 0.04 mg/l instead of an approx. SC200. An injection of 10 mg vitamin B2 would instantaneously result in a plasma concentration of about 3 mg/l vitamin B2. This is 30fold the mean value of the approx. SC200 on AM-coagulation in 5 of 6 samples. Therefore, vitamin B2 should not be administered intravenously, oral vitamin B2 has a much slower increase of plasma concentration. If there are indeed clinical situations where vitamin B2 injection is necessary, thiamine should be accompanied by low-molecular-weight-heparin (LMWH) thrombosis prophylaxis.

INTRODUCTION

Riboflavine is the central component of the cofactors flavin adenine dinucleotide (FAD) and flavin mononucleotide (FMN). FAD and FMN are bound coenzymes of many redox active enzymes. FAD or (FMN) accept 2 hydrognes, yielding FADH$_2$ (or FMNH$_2$). Flavoenzymes decarboxylate pyruvate and 2-oxo-glutarate, dehydrogenate fatty acyl CoA in fatty acid ß-oxidation (3-oxidation), activate (oxidize) vitamin B$_6$ (pyridoxal phosphate) and retinol (vitamin A), synthesize the active form of folate (5-methyl THF), reduce oxidized glutathione (GSSG), reduce cytochrome P450, oxidize NADPH to H_2O_2. Flavoproteins participate in the electron transport chain of mitochondrial respiration [1-4].

The daily requirement of riboflavin is about 1-2 mg [1]. The normal concentration in blood is 75-300 µg/l; deficiencies of riboflavin result in inflammation of skin and mucous membranes, mouth tears, itching, anemia, cataracts, retinal detachment, atherothrombosis [5].

Riboflavine is clinically available in 10 mg amounts per ampoule. High dose riboflavin (400 mg per day) is indicated in certain migraine forms [6-9]. 50 µmol/l (18.4 mg/l) riboflavin is used in UV-illuminated riboflavin-photosensitized pathogen-inactivation of donor blood [10], but 0.5-1 mM chloramine treatment is superior [11-15].

Since riboflavine contains 2 delta-negatively charged oxygens in keto-position, 4 delta-negatively charged oxygens in hydroxyl-groups, and a lipophilic benzene ring with 2 methyl-groups, riboflavin could trigger altered matrix (AM) - coagulation; negatively charged or lipophilic molecules fold factor 12 into F12a or pre-kallikrein into kallikrein [16-19]. The best assay to determine even spurious activations of AM-coagulation is the recalcified coagulation activity assay (RECA) [20].

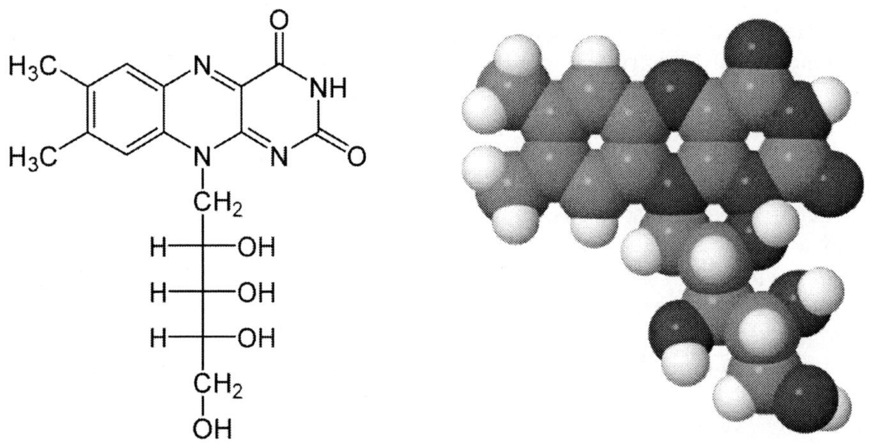

Figure 1. Chemical structure of vitamin B_2 (riboflavine) [1].
Riboflavine (6,7-dimethyl-9-D-ribitylisoalloxazine; $C_{17}H_{20}N_4O_6$; 367,4 Daltons) is oxidized pteridine (isoalloxazine) coupled to ribitol. The lipophilic parts of the molecule are in grey, the delta-negatively charged ones in red. Pyrimidine linked via methylene to thiazole ($C_{12}H_{18}ClN_4OS$; thiamin · HCl; molecular weight 332.3). Ester reaction of thiamine's hydroxyl group with activated pyro-phosphoric acid (in adenosine triphosphate) transforms thiamine into the active metabolite thiamine diphosphate (TPP).

MATERIAL AND METHODS

40 µl -30°C frozen/23°C thawed platelet poor citrated plasma of 6 healthy donors after written informed consent were supplemented with 0-3.05 mg/l riboflavin (Streuli, Uznach, Switzerland; 10 mg amount in 2 ml ampoules) by repetitive 1+1 dilution on the polystyrene microtiter plate (Brand, Wertheim, Germany; article nr. 781600). The RECA was started with 4 µl 250 mM $CaCl_2$ (Sigma, Deisenhofen, Germany). After 0, 10, or 20 min (37°C) coagulation reaction time (CRT) the RECA was stopped by addition of 80 µl 2.5 M arginine,

pH 8.6, 0.16% Triton X 100® (Sigma). After 3 min 20 µl chromogenic thrombin substrate 1 mM HD-CHG-Ala-Arg-pNA (Pentapharm, Basel, Switzerland) in 1.25 M arginine, pH 8.7, were added and the increase in absorbance with time $\Delta A/t$ was determined at 405 nm by a microtiter plate photometer with a 1 mA resolution (PHOmo; Autobio-anthos, Krefeld, Germany). The approximate 200% stimulatory concentration (approx. SC200) or the approximate 50% inhibitory concentration (approx. IC50) was determined for each individual plasma at the most sensible thrombin generation curve. Considered were only the thrombin activities in the ascending part of the CRT vs. thrombin activity generation curve.

RESULTS AND DISCUSSION

Figures 2-7 demonstrate that the approx. SC200 on AM-coagulation was 0.1±0.04 mg/l (MV±1SD) vitamin B$_2$ (range: 0.04-1.5 mg/l) in 5 of 6 samples. The 6th sample had an approx. IC50 of 0.04 mg/l riboflavin instead of an approx. SC200. An injection of 10 mg vitamin B$_2$ would instantaneously result in a plasma concentration of about 3 mg/l vitamin B$_2$. This is 30fold the mean value of the approx. SC200 on AM-coagulation in 5 of 6 samples. Therefore, vitamin B2 should not be administered intravenously, oral vitamin B$_2$ has a much slower increase of plasma concentration. If there are indeed clinical situations where vitamin B$_2$ injection is necessary, thiamine should be accompanied by low-molecular-weight-heparin (LMWH) thrombosis prophylaxis [21].

The question arises why are there certain plasmas that respond to vitamin B$_2$ addition with an IC50 instead of a SC200. Here riboflavin seems to stop AM-coagulation by blocking protein/protein interactions. Riboflavin could attach to certain important proteins of intrinsic coagulation and act as an AM-coagulation interrupter. Clarifying the exact mechanism of AM-coagulation interruption would be of great pharmacologic interest.

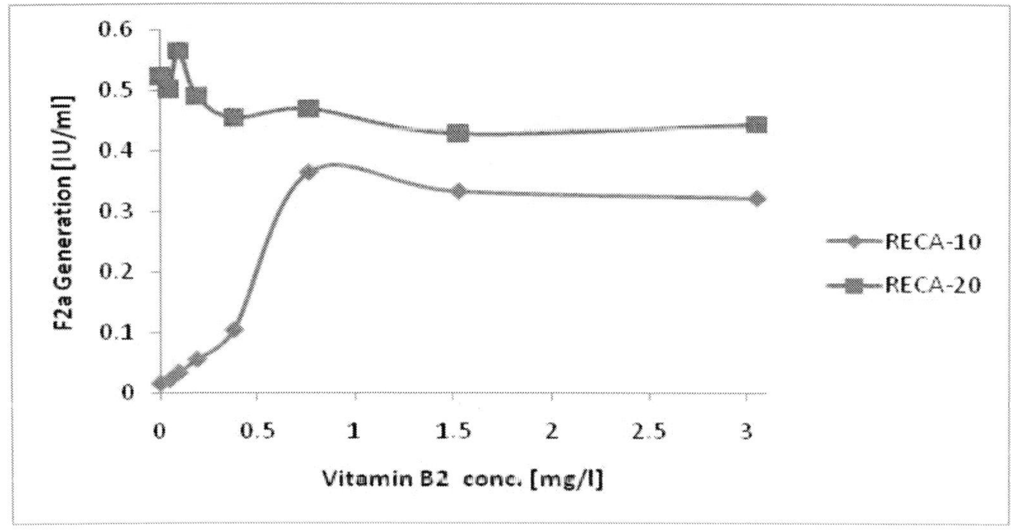

Figure 2. SC200 determination in plasma 1. 40 µl citrated plasma 1, supplemented with 0-3.05 mg/l vitamin B$_2$, were recalcified and the thrombin generation was stopped after 10 min or 20 min by arginine as described under Methods. The approx. SC200 was 0.1 mg/l.

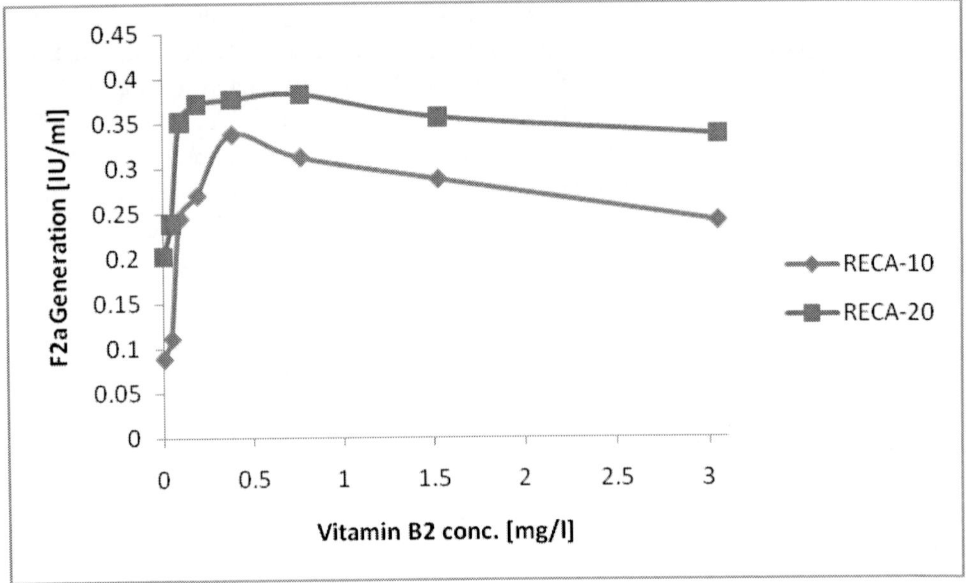

Figure 3. SC200 determination in plasma 2. 40 µl citrated plasma 2, supplemented with 0-3.05 mg/l vitamin B₂, were recalcified and the thrombin generation was stopped after 10 min or 20 min by arginine as described under Methods. The approx. SC200 was 0.1 mg/l.

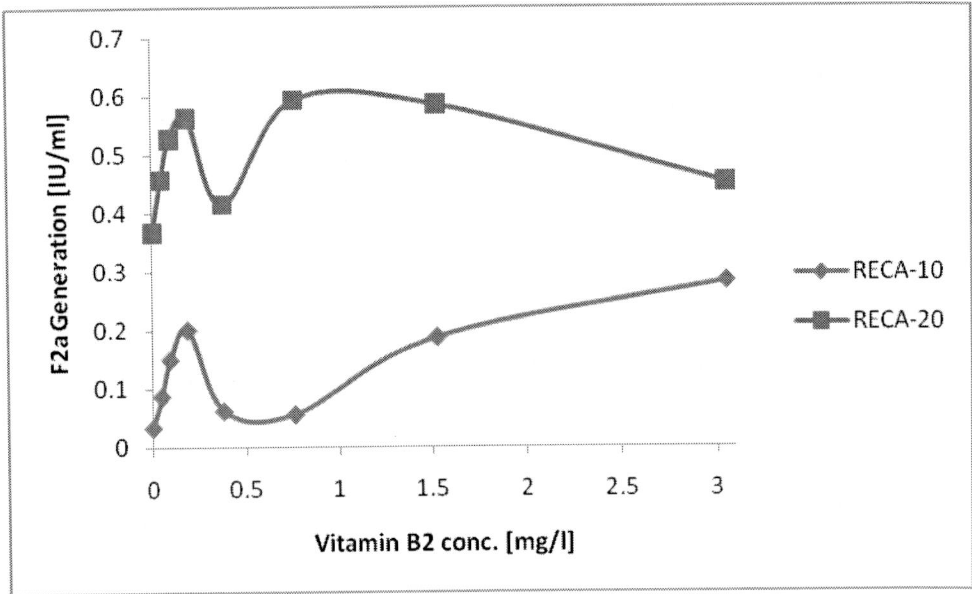

Figure 4. SC200 determination in plasma 3. 40 µl citrated plasma 3, supplemented with 0-3.05 mg/l vitamin B₂, were recalcified and the thrombin generation was stopped after 10 min or 20 min by arginine as described under Methods. The approx. SC200 was 0.04 mg/l.

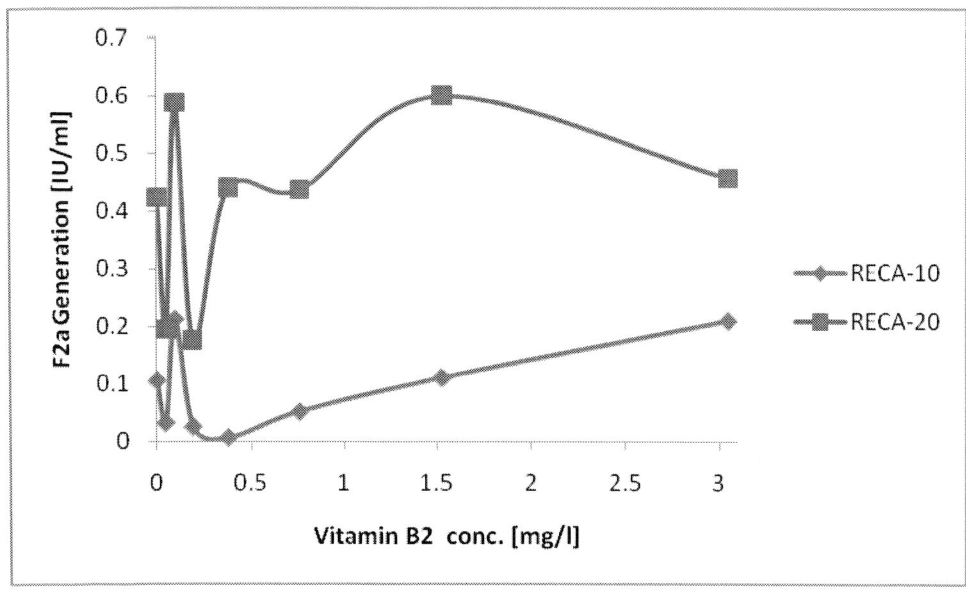

Figure 5. SC200 determination in plasma 4. 40 µl citrated plasma 4, supplemented with 0-3.05 mg/l vitamin B$_2$, were recalcified and the thrombin generation was stopped after 10 min or 20 min by arginine as described under Methods. The approx. SC200 was 0.1 mg/l.

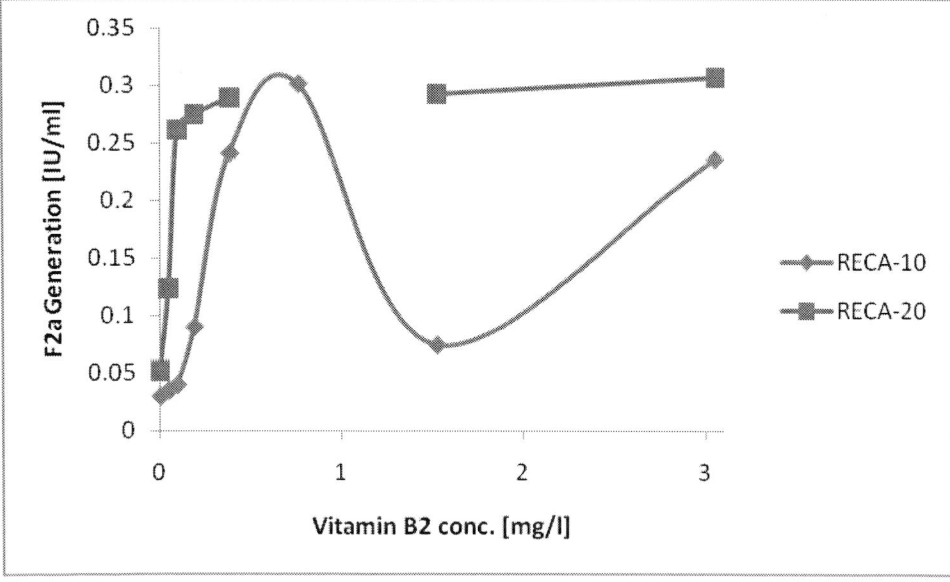

Figure 6. SC200 determination in plasma 5. 40 µl citrated plasma 5, supplemented with 0-3.05 mg/l vitamin B$_2$, were recalcified and the thrombin generation was stopped after 10 min or 20 min by arginine as described under Methods. The approx. SC200 was 0.15 mg/l.

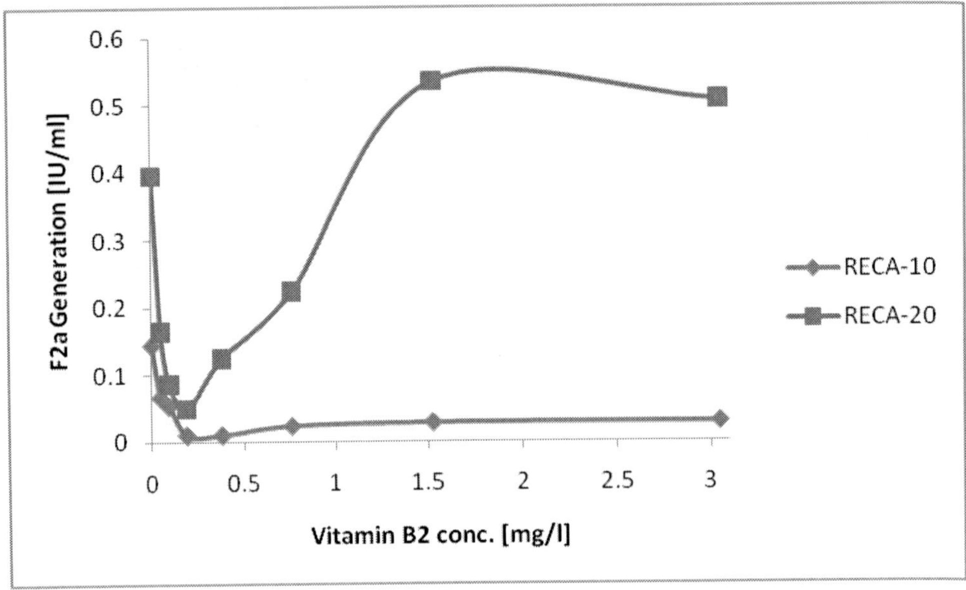

Figure 7. IC50 determination in plasma 6. 40 µl citrated plasma 6, supplemented with 0-3.05 mg/l vitamin B$_2$, were recalcified and the thrombin generation was stopped after 10 min or 20 min by arginine as described under Methods. The approx. IC50 was 0.04 mg/l.

REFERENCES

[1]   www.wikipedia.org

[2]   Walsh CT, Wencewicz TA. Flavoenzymes: versatile catalysts in biosynthetic pathways. Nat Prod Rep. 2013; 30: 175-200.

[3]   Pudney CR, Heyes DJ, Khara B, Hay S, Rigby SE, Scrutton NS. Kinetic and spectroscopic probes of motions and catalysis in the cytochrome P450 reductase family of enzymes. FEBS J. 2012; 279: 1534-44.

[4]   Stief TW. Neutrophil granulocytes in hemostasis. *Hemostasis Laboratory* 2008; 1: 269-89.

[5]   www.aok.de

[6]   Krymchantowski AV, Bigal ME, Moreira PF. New and emerging prophylactic agents for migraine. CNS Drugs. 2002; 16: 611-34.

[7]   Markley HG. CoEnzyme Q10 and riboflavin: the mitochondrial connection. Headache. 2012; 52 Suppl 2: 81-7.

[8]   Schoenen J, Jacquy J, Lenaerts M. Effectiveness of high-dose riboflavin in migraine prophylaxis. A randomized controlled trial. *Neurology.* 1998; 50: 466-70.

[9]   Boehnke C, Reuter U, Flach U, Schuh-Hofer S, Einhäupl KM, Arnold G. High-dose riboflavin treatment is efficacious in migraine prophylaxis: an open study in a tertiary care centre. *Eur J Neurol.* 2004; 11: 475-7.

[10]   Marschner S, Goodrich R. Pathogen reduction technology treatment of platelets, plasma and whole blood using riboflavin and UV light. *Transfus Med Hemother.* 2011; 38: 8-18.

[11] Stief TW, Kurz J, Doss MO, Fareed J. Singlet oxygen (^{1}O$_2$) inactivates fibrinogen, factor V, factor VIII, factor X, and platelet aggregation of human blood. *Thromb Res* 2000; 97: 473-80.

[12] Stief TW, Slenczka W, Renz H, Klenk HD. Singlet oxygen (^{1}O$_2$) generating chloramines at concentrations that are tolerable for normal hemostasis function inactivate the lipid enveloped vesicular stomatitis virus in human blood. In: 3rd Symposion on the Biology of Endothelial Cells; Pathophysiology of the Endothelium: Vascular and Infectious Diseases, May 24-26, 2001, Giessen, Germany, Abstr. D10.

[13] Stief TW. Hemostasis tolerable singlet oxygen - a perspective in AIDS therapy. *Hemostasis Laboratory* 2008; 1: 21-40.

[14] Stief TW. Singlet oxygen and thrombin generation in RECA. *Hemostasis Laboratory* 2011; 4: 237-254.

[15] Stief TW. Singlet oxygen and thrombin generation: 0.5-1 mM chloramine as anti-viral therapy. *Hemostasis Laboratory* 2010; 3: 311-24.

[16] Stief TW. Factor 12 activation in two purified systems. *Hemostasis Laboratory* 2011; 4: 385-408.

[17] Stief TW. Zn^{2+}, hexane, or glucose activate factor 12 and/or prekallikrein in two purified systems. *Hemostasis Laboratory* 2011; 4: 409-426.

[18] Stief TW. Zn^{2+}, hexane, valproate, or glucose in two purified systems of F12-PK-HMWK. *Hemostasis Laboratory* 2012; 5: 35-50.

[19] Stief TW, Mohrez M. HMWK increases of decreases F12a generation dependent on the contact trigger concentration in two purified systems. *Hemostasis Laboratory* 2012; 5: 51-65.

[20] Stief TW. Pathological thrombin generation by the synthetic inhibitor argatroban. *Hemostasis Laboratory* 2009; 2: 83-104.

[21] Stief TW. LMWH - action-monitoring for all patients. Acta Paediatr. 2012; 101: e314.

Chapter 8

VITAMIN B₆ TRIGGERS AM-COAGULATION

ABSTRACT

Background: Vitamin B_6 in its molecular forms pyridoxine, pyridoxal, pyridoxamine, and in its active form pyridoxal phosphate is an important cofactor in many reactions of amino acid metabolism. Pyridoxine with its 3 delta-negatively charged oxygens could trigger intrinsic coagulation.

Material and Methods: 40 µl platelet poor citrated plasma of 6 healthy donors were supplemented with 0-30.5 mg/l pyridoxine hydrochloride by repetitive 1+1 dilution on the polystyrene microtiter plate (Brand®781600). The RECA was performed with 10 or 20 min (37°C) coagulation reaction time. The approx. SC200 or approx. IC50 was determined for each individual plasma.

Results and Discussion: The approx. SC200 on AM-coagulation was 1.6±1.0 mg/l (MV±1SD) pyridoxine in 4 of 6 samples. 2 of 6 samples had an approx. IC50 of 0.7 mg/l, in these 2 plasmas vitamin B6 seems to stop AM-coagulation by blocking protein/protein interactions acting like an AM-coagulation interrupter. Vitamin B_6 should not be administered intravenously, oral vitamin B6 has a much slower increase of plasma concentration.

INTRODUCTION

Vitamin B_6 occurs in the 3 different molecular forms pyridoxine, pyridoxal, pyridoxamine with their respective phosphorylated states [1]. Pyridoxal phosphate (generated by pyridoxal kinase) is the active form and is a cofactor in many reactions of amino acid metabolism, including transamination, deamination, and decarboxylation, generating the active amines histamine (out of histidine), serotonin (out of tryptophan), gamma-aminobutyric acid (out of glutamate), and dopamine out of dihydroxyphenylalanine [1].

Pyridoxal phosphate is also a cofactor of serine racemase that synthesizes the neuromodulator D-serine (normally all human amino acids are L-amino acids). Vitamin B_6 acts in DNA or RNA metabolism in still not completely understood pathways [2]. Vitamin B_6 inhibits the transcription of glucocorticoid synthesizing enzymes, and modulates the expression of albumin mRNA and of glycoprotein IIb (the fibrinogen receptor of thrombocytes) [1]. Thus, vitamin B_6 is necessary for normal hemostasis function [3].

The daily requirement of vitamin B_6 is about 1-2 mg. the normal concentration in EDTA-blood is 5-30 µg/l [4]. 4-Pyridoxic acid is its inactive metabolite that is renally eliminated.

Vitamin B_6 deficiency results in dermatitis seborrhoica, atrophic glossitis, confusion, and neuropathy [1].

Vitamin B6 is clinically available in 100 mg pyridoxine per ampoule. Since pyridoxine contains 3 delta-negatively charged oxygens linked to CH_2-C injected vitamin B6 could strongly trigger altered matrix (AM) coagulation; negatively charged or lipophilic molecules fold factor 12 into F12a or pre-kallikrein into kallikrein [5-10]. The best assay to determine even spurious activations of AM-coagulation is the recalcified coagulation activity assay (RECA) [11].

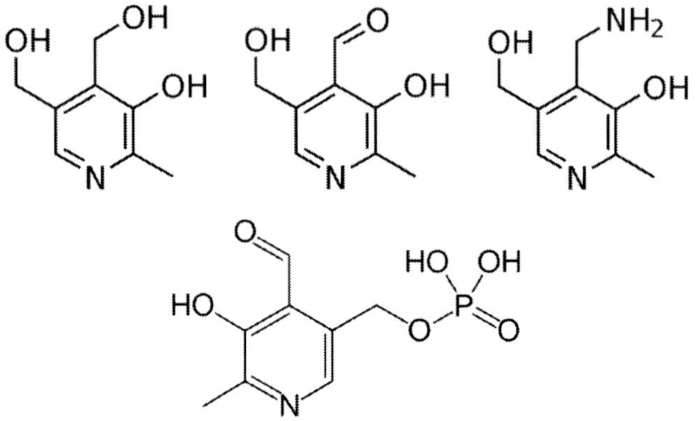

Figure 1. Chemical structure of vitamin B_6 [1]. Vitamin B_6 occurs in the 3 forms pyridoxine (left), pyridoxal (middle), and pyridoxamine (right). Active vitamin B_6 is phosphorylated to pyridoxal phosphate (below). 3-Hydroxy-5-(hydroxymethyl)-2-methyl-4-pyridinecarboxaldehyde hydrochloride (pyridoxal) has 203.6 Daltons.

MATERIAL AND METHODS

40 µl -30°C frozen/23°C thawed platelet poor citrated plasma of 6 healthy donors that gave written informed consent were supplemented with 0-30.5 mg/l pyridoxine hydrochloride (ratiopharm, Ulm, Germany; 2 ml ampoules of 50 mg/ml) by repetitive 1+1 dilution on the polystyrene microtiter plate (Brand, Wertheim, Germany; article nr. 781600). The RECA was started with 4 µl 250 mM $CaCl_2$. After 0, 10, or 20 min (37°C) coagulation reaction time (CRT) the RECA was stopped by addition of 80 µl 2.5 M arginine, pH 8.6, 0.16% Triton X 100® (Sigma). After 3 min 20 µl chromogenic thrombin substrate 1 mM HD-CHG-Ala-Arg-pNA (Pentapharm, Basel, Switzerland) in 1.25 M arginine, pH 8.7, were added and the increase in absorbance with time $\Delta A/t$ was determined at 405 nm by a microtiter plate photometer with a 1 mA resolution (PHOmo; Autobio-anthos, Krefeld, Germany). The approximate 200% stimulatory concentration (approx. SC200) or the approximate 50% inhibitory concentration (approx. IC50) was determined for each individual plasma at the most sensible thrombin generation curve. Considered were only the thrombin activities in the ascending part of the CRT vs. thrombin activity generation curve.

RESULTS AND DISCUSSION

Figures 2-7 demonstrate that the approx. SC200 on AM-coagulation was 1.6±1.0 mg/l (MV±1SD) vitamin B$_6$ (range: 0.5-1.5 mg/l) in 4 of 6 samples. 2 of 6 samples had an approx. IC50 of 0.7 mg/l, in these 2 plasmas vitamin B$_6$ seems to stop AM-coagulation by blocking protein/protein interactions acting like an AM-coagulation interrupter.

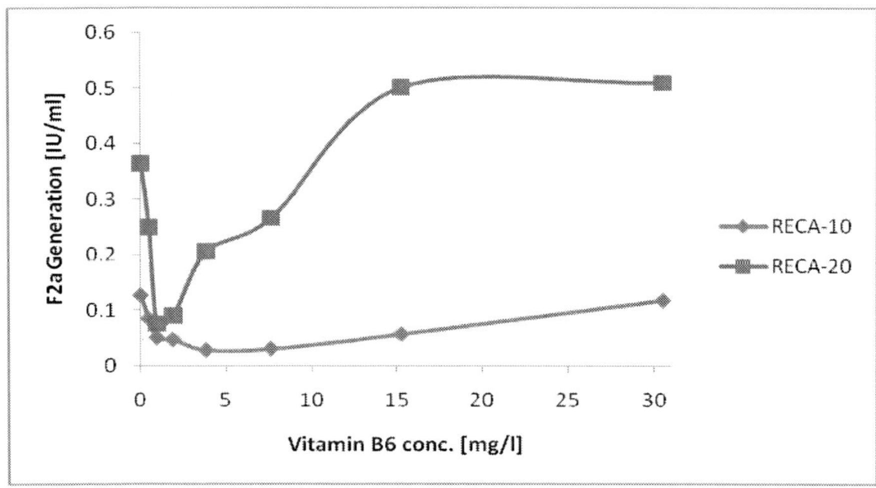

Figure 2. SC200 determination in plasma 1. 40 µl citrated plasma 1, supplemented with 0-30.5 mg/l vitamin B6, were recalcified and the thrombin generation was stopped after 10 min or 20 min by arginine as described under Methods. The approx. IC50 was 0.7 mg/l.

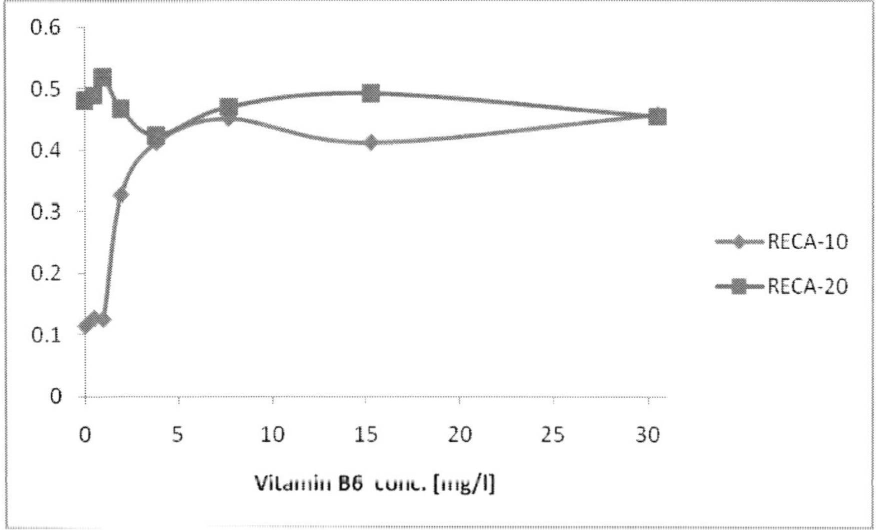

Figure 3. SC200 determination in plasma 2. 40 µl citrated plasma 2, supplemented with 0-30.5 mg/l vitamin B6, were recalcified and the thrombin generation was stopped after 10 min or 20 min by arginine as described under Methods. The approx. SC200 was 1.5 mg/l.

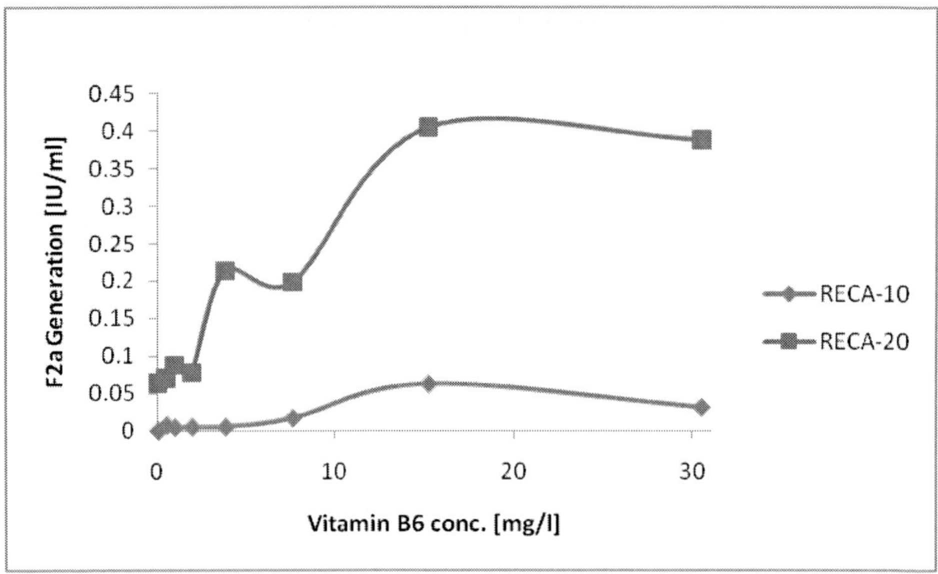

Figure 4. SC200 determination in plasma 3. 40 µl citrated plasma 3, supplemented with 0-30.5 mg/l vitamin B6, were recalcified and the thrombin generation was stopped after 10 min or 20 min by arginine as described under Methods. The approx. SC200 was 3 mg/l.

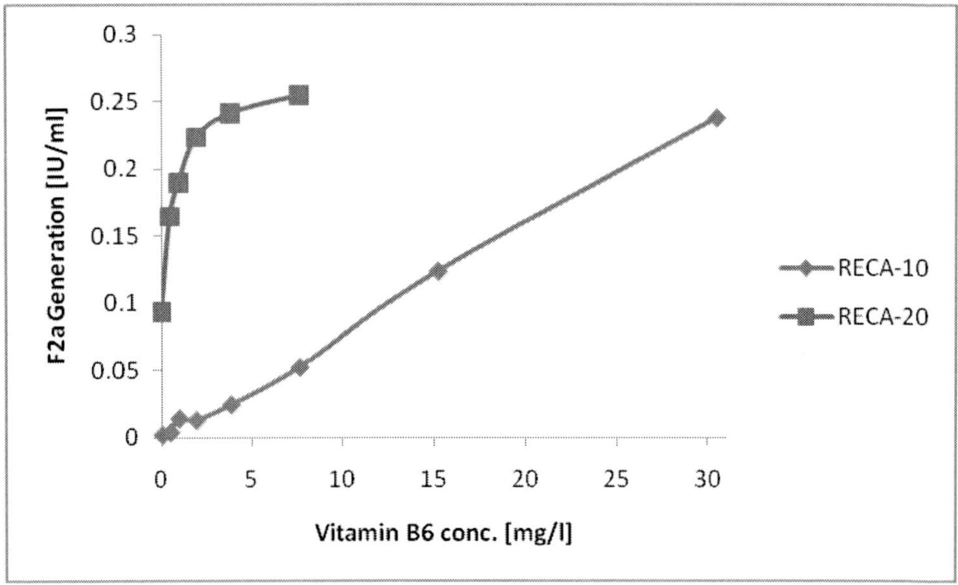

Figure 5. SC200 determination in plasma 4. 40 µl citrated plasma 4, supplemented with 0-30.5 mg/l vitamin B6, were recalcified and the thrombin generation was stopped after 10 min or 20 min by arginine as described under Methods. The approx. SC200 was 0.5 mg/l.

An injection of 100 mg vitamin B_6 would instantaneously result in a plasma concentration of about 30 mg/l vitamin B_6. This is about 20fold the mean value of the approx. SC200 on AM-coagulation in 4 of 6 samples. Vitamin B6 deficiency could result in thrombosis [3], however too much vitamin B_6 could also generate increased concentrations of circulating micro-thrombi. Therefore, vitamin B_6 should not be administered intravenously, oral vitamin B$_6$ has a much slower increase of plasma concentration. If there are really clinical situations where vitamin B$_6$ injection is necessary, the drug should be accompanied by low-molecular-weight-heparin (LMWH) thrombosis prophylaxis [12].

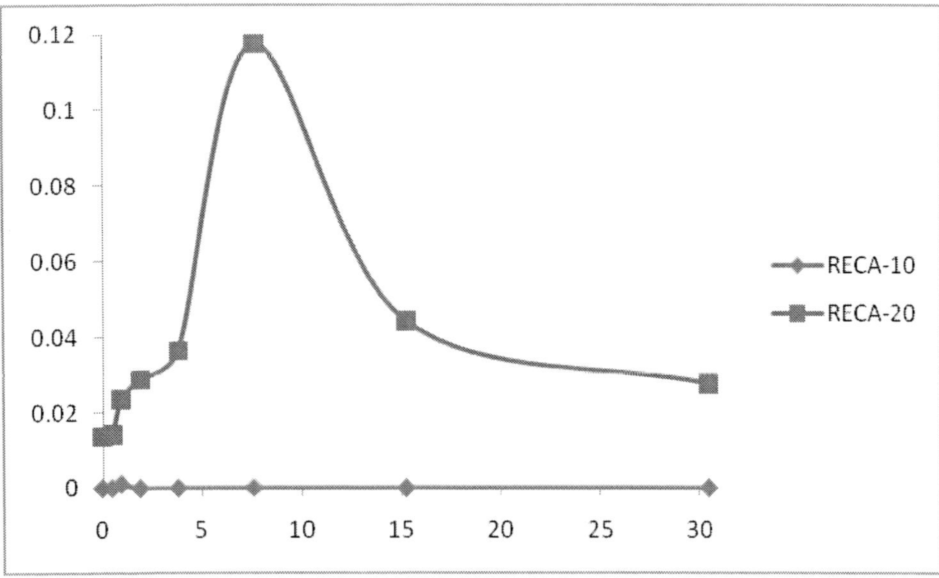

Figure 6. SC200 determination in plasma 5. 40 μl citrated plasma 5, supplemented with 0-30.5 mg/l vitamin B6, were recalcified and the thrombin generation was stopped after 10 min or 20 min by arginine as described under Methods. The approx. SC200 was 1.5 mg/l.

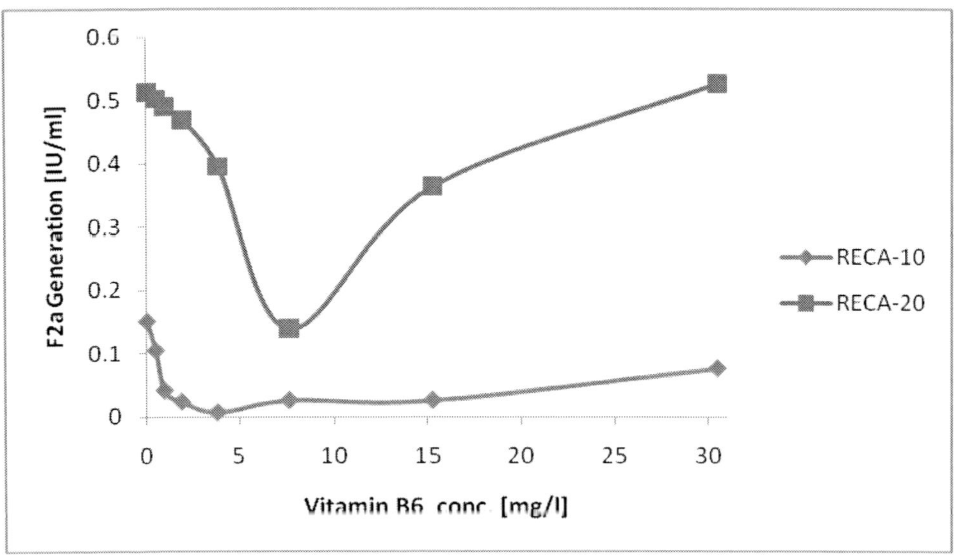

Figure 7. SC200 determination in plasma 6. 40 μl citrated plasma 6, supplemented with 0-30.5 mg/l vitamin B6, were recalcified and the thrombin generation was stopped after 10 min or 20 min by arginine as described under Methods. The approx. IC50 was 0.7 mg/l.

REFERENCES

[1] www.wikipedia.org

[2] Galluzzi L, Marsili S, Vitale I, Senovilla L, Michels J, Garcia P, Vacchelli E, Chatelut E, Castedo M, Kroemer G. Vitamin B6 metabolism influences the intracellular accumulation of cisplatin. Cell Cycle. 2013; 12: 417-21.

[3] Zhou K, Zhao R, Geng Z, Jiang L, Cao Y, Xu D, Liu Y, Huang L, Zhou J. Association between B-group vitamins and venous thrombosis: systematic review and meta-analysis of epidemiological studies. J Thromb Thrombolysis. 2012; 34: 459-67.

[4] www.aok.de

[5] Stief TW. The contact phase activity assay. *Hemostasis Laboratory* 2009; 1: 1-22.

[6] Stief TW. Factor 12 activation in two purified systems. *Hemostasis Laboratory* 2011; 4: 385-408.

[7] Stief TW. Zn^{2+}, hexane, or glucose activate factor 12 and/or prekallikrein in two purified systems. *Hemostasis Laboratory* 2011; 4: 409-426.

[8] Stief TW. Zn^{2+}, hexane, valproate, or glucose in two purified systems of F12-PK-HMWK. *Hemostasis Laboratory* 2012; 5: 35-50.

[9] Stief TW, Mohrez M. HMWK increases of decreases F12a generation dependent on the contact trigger concentration in two purified systems. *Hemostasis Laboratory* 2012; 5: 51-65.

[10] Stief TW. Coumarins trigger altered matrix (AM) coagulation activation. *Hemostasis Laboratory* 2013; 6: 121-8.

[11] Stief TW. Pathological thrombin generation by the synthetic inhibitor argatroban. *Hemostasis Laboratory* 2009; 2: 83-104.

[12] Stief TW. LMWH - action-monitoring for all patients. Acta Paediatr. 2012; 101: e314.

Chapter 9

VITAMIN B$_{12}$ MODULATES AM-COAGULATION

ABSTRACT

Background: Cobalamin is the hydrophilic vitamin B$_{12}$ of a complicated structure with a central cobalt ion for normal erythropoiesis and nerve function. Vitamin B$_{12}$ can clinically be given as 1000-1500 µg injections. Therefore, this work quantified the changed thrombin generation by alteration of matrix thru an increase of vitamin B$_{12}$ concentration.

Material and Methods: 40 µl platelet poor citrated plasma of 12 healthy donors were supplemented either with 0-610 µg/l hydroxocobalamin acetate or with 0-610 µg/l cyanocobalamine by repetitive 1+1 dilution on the polystyrene microtiter plate (Brand®781600). The RECA was performed with 10 or 20 min (37°C) coagulation reaction time. The approx. SC200 or approx. IC50 was determined for each individual plasma.

Results and Discussion: The approx. SC200 on AM-coagulation was 19±14 µg/l (MV±1SD) hydroxocobalamin acetate in 5 of 6 plasmas; one plasma had an approx. IC50 of 8 µg/l. Cyanocobalamin had an approx. SC200 of 101±130 µg/l in 3 of 6 plasmas and an approx. IC50 of 13±14 µg/l in 3 of 6 samples. This means that there are hemostasis differences between hydroxocobalamin acetate and cyanocobalamin. Hydroxocobalamin acetate triggers intrinsic coagulation much more than cyanocobalamin. This might be due to the acetate part of the hydroxocobalamin drug. It could be of pharmacologic interest to substitute acetate in hydroxocobalamin by HCl.

INTRODUCTION

Vitamin B$_{12}$ with its 4 cobalamin vitamers adenosyl-, methyl-, cyano-, and hydroxo- is a C$_1$ transfering with folate cooperating vitamin, necessary for the generation of erythrocytes and for normal nerve function [1-4]. The drug vitamin B$_{12}$ acts against cobalamin deficiency, transcobalamin deficienciy, and cyanide poisoning. For the latter indication, a large amount of hydroxocobalamine is given intravenously; cyanide converts hydroxo- in cyano-cobalamine that is excreted in urine.

The daily requirement of vitamin B$_{12}$ is 2 to 3 µg [1]. The normal plasma concentration is about 200-900 ng/l [5].

Vitamin B_{12} is clinically available in 1-1.5 mg amounts per ampoule. Since cobalamin is of such a complicated molecular structure (Figure 1) injection of cobalamin could induce a blood matrix change, causing altered matrix (AM)-coagulation activation [6,7]; negatively charged or lipophilic molecules fold factor 12 into F12a or pre-kallikrein into kallikrein [8-11]. The best assay to determine even spurious activations of AM-coagulation is the recalcified coagulation activity assay (RECA) [12].

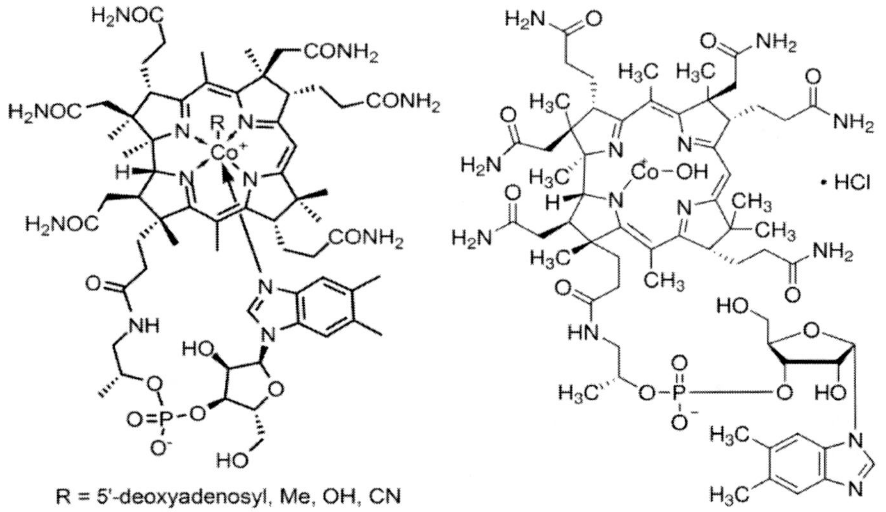

R = 5'-deoxyadenosyl, Me, OH, CN

Figure 1. Chemical structure of vitamin B_{12} (cobalamin). Vitamin B_{12} in its 4 vitamer forms adenosylcobalamin, methylcobalamin, hydroxocobalamin, and cyanocobalamin (left) [1]. The hydroxocobalamin·HCl form (right) [2] has the sum formula $C_{62}H_{89}CoN_{13}O_{15}P$·HCl with a molecular mass of 1382.8. Vitamin B_{12} is absorbed together with intrinsic factor in the distal half of the ileum. In blood it is transported by transcobalamins, hydroxocobalamin binds slightly higher than cyanocobalamin. Its half-live in blood is about 6d, in the liver about 400d. Vitamin B_{12} is renally eliminated.

MATERIAL AND METHODS

40 µl -30°C frozen/23°C thawed platelet poor citrated plasma of 12 healthy donors were supplemented after written informed consent either with 0-609.8 µg/l hydroxocobalamin acetate (Pascoe, Giessen, Germany; 1 ml ampoules of 1500 µg) or with cyanocobalamin (Jenapharm, Brehna, Germany; 1 ml ampoules of 1000 µg) by repetitive 1+1 dilution on the polystyrene microtiter plate (Brand, Wertheim, Germany; article nr. 781600). The RECA was started with 4 µl 250 mM $CaCl_2$ (Sigma, Deisenhofen, Germany) After 0, 10, or 20 min (37°C) coagulation reaction time (CRT) the RECA was stopped by addition of 80 µl 2.5 M arginine, pH 8.6, 0.16% Triton X 100® (Sigma). After 3 min 20 µl chromogenic thrombin substrate 1 mM HD-CHG-Ala-Arg-pNA (Pentapharm, Basel, Switzerland) in 1.25 M arginine, pH 8.7, were added and the increase in absorbance with time $\Delta A/t$ was determined at 405 nm by a microtiter plate photometer with a 1 mA resolution (PHOmo; Autobio-anthos, Krefeld, Germany). The approximate 200% stimulatory concentrations (approx. SC200) or the approximate 50% inhibitory concentrations (approx. IC50) were determined for the

individual plasmas at the most sensible thrombin generation curve. Considered were only the thrombin activities in the ascending part of the CRT vs. thrombin activity generation curve.

RESULTS AND DISCUSSION

Figures 2-7 demonstrate the results of hydroxocobalamin acetate, figures 8-13 that of cyanocobalamin. The approx. SC200 on AM-coagulation was 19±14 µg/l (MV±1SD) hydroxocobalamin acetate in 5 of 6 plasmas; one plasma had an approx. IC50 of 8 µg/l (Figure 6). Cyanocobalamin had an approx. SC200 of 101±130 µg/l in 3 of 6 plasmas and an approx. IC50 of 13±14 µg/l in 3 of 6 samples. This means that there are pronounced hemostasis differences between hydroxocobalamin acetate and cyanocobalamin. Hydroxocobalamin acetate triggers intrinsic coagulation more than cyanocobalamin. This might be due to the acetate part of the hydroxocobalamin drug. Acetate triggers intrinsic coagulation with approx. SC200 values in the range of about 10 µM [13]. It could be of pharmacologic interest to substitute acetate in hydroxocobalamin by HCl. Cobalamin injection in hemostasis susceptible patients [14,15] might be accompanied by low-molecular-weight-heparin (LMWH) thrombosis prophylaxis [16]. The question arises why there are plasmas that respond to vitamin B$_{12}$ addition with an IC50 instead of a SC200. Here cobalamin seems to stop AM-coagulation by blocking protein/protein interactions. In certain plasmas cobalamin could attach to certain important proteins of intrinsic coagulation and act as an AM-coagulation interrupter. Clarifying the exact mechanism of AM-coagulation interruption would be of great pharmacologic interest.

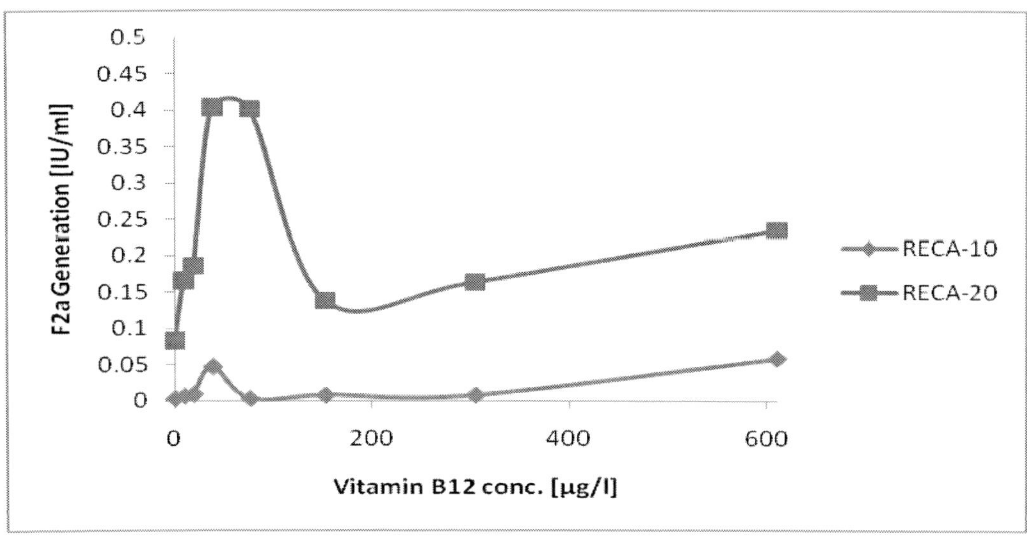

Figure 2. SC200 determination in plasma 1. 40 µl citrated plasma 1, supplemented with 0-610 µg/l hydroxocobalamin acetate, were recalcified and the thrombin generation was stopped after 10 min or 20 min by arginine as described under Methods. The approx. SC200 was 9 µg/l.

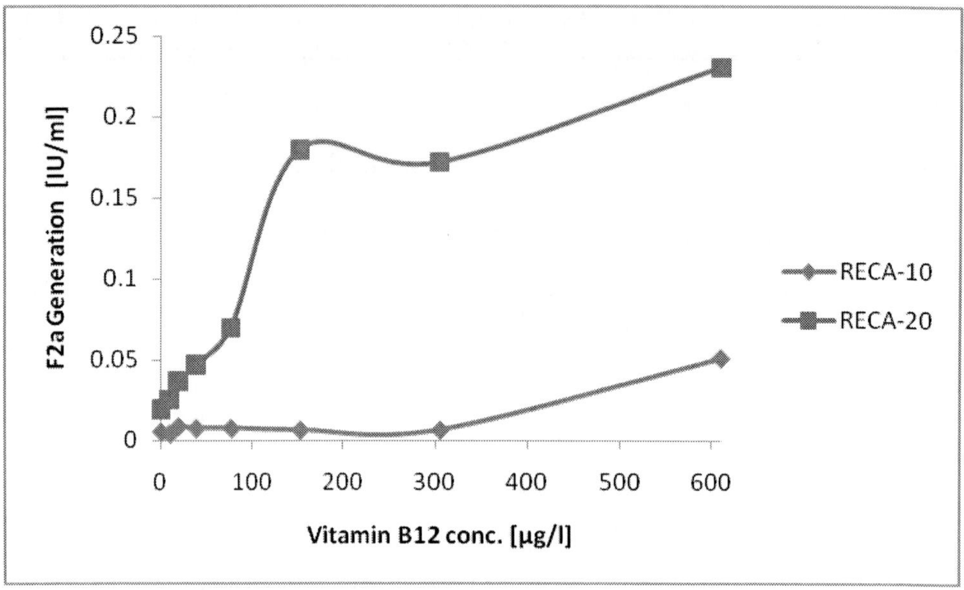

Figure 3. SC200 determination in plasma 2. 40 µl citrated plasma 2, supplemented with 0-610 µg/l hydroxocobalamin acetate, were recalcified and the thrombin generation was stopped after 10 min or 20 min by arginine as described under Methods. The approx. SC200 was 20 µg/l.

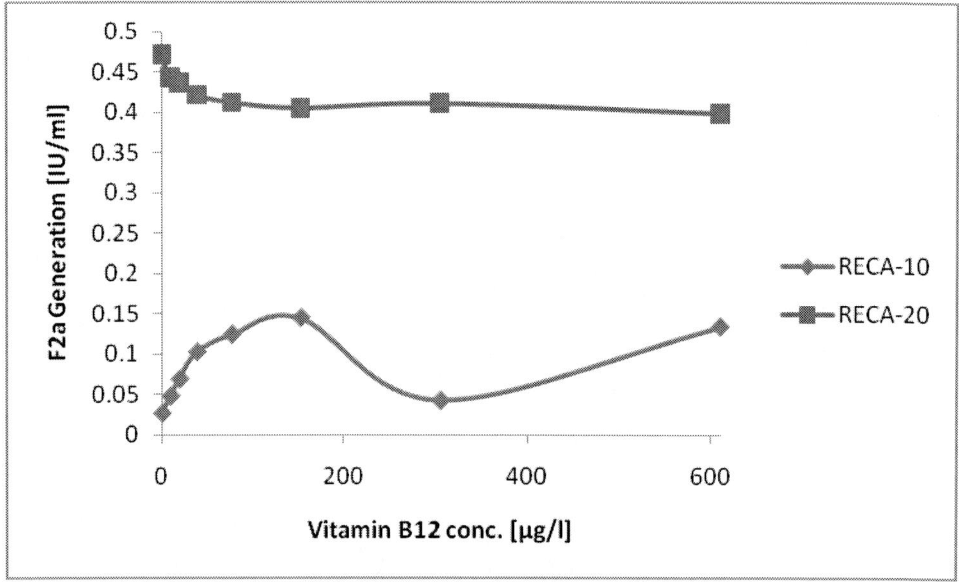

Figure 4. SC200 determination in plasma 3. 40 µl citrated plasma 3, supplemented with 0-610 µg/l hydroxocobalamin acetate, were recalcified and the thrombin generation was stopped after 10 min or 20 min by arginine as described under Methods. The approx. SC200 was 4 µg/l.

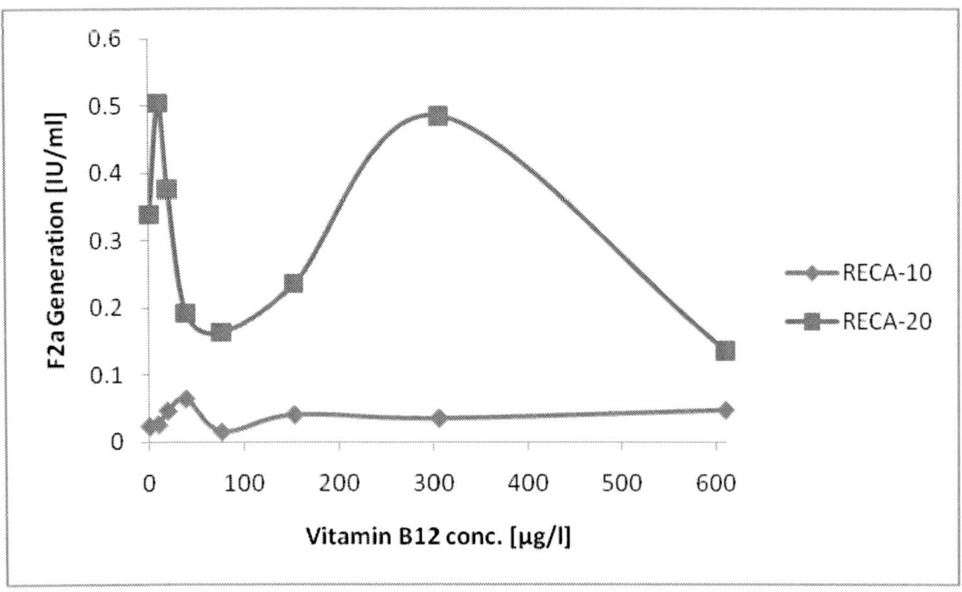

Figure 5. SC200 determination in plasma 4. 40 µl citrated plasma 4, supplemented with 0-610 µg/l hydroxocobalamin acetate, were recalcified and the thrombin generation was stopped after 10 min or 20 min by arginine as described under Methods. The approx. SC200 was 20 µg/l.

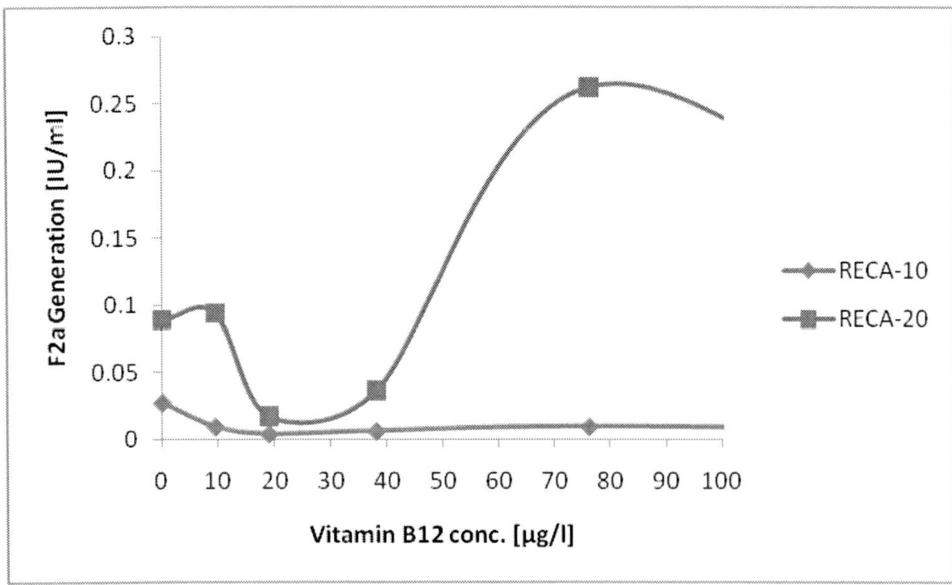

Figure 6. IC50, SC200 determination in plasma 5. 40 µl citrated plasma 5, supplemented with 0-610 µg/l hydroxocobalamin acetate, were recalcified and the thrombin generation was stopped after 10 min or 20 min by arginine as described under Methods. The approx. IC50 was 8 µg/l, the approx. SC200 was 55 µg/l.

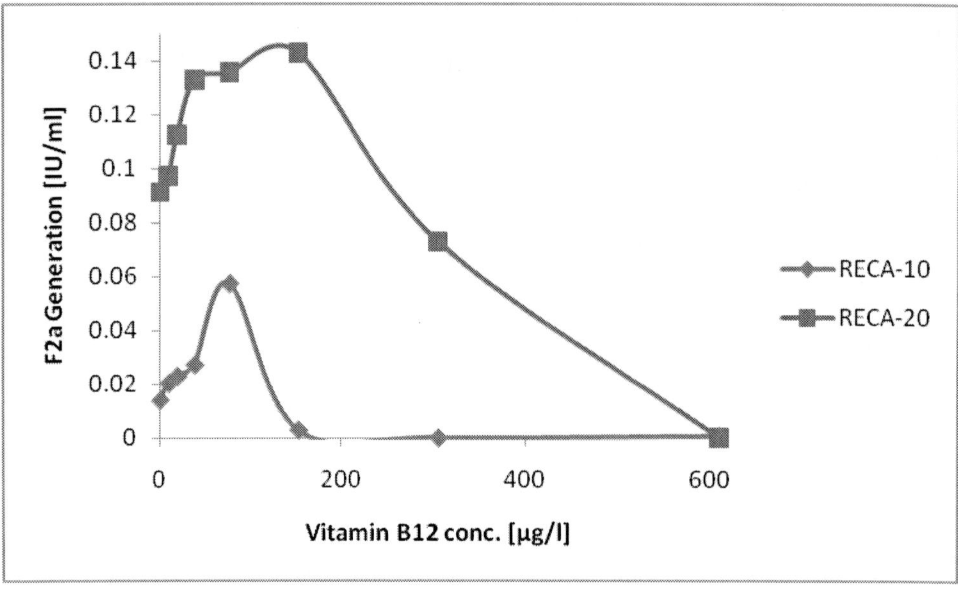

Figure 7. SC200 determination in plasma 6. 40 µl citrated plasma 6, supplemented with 0-610 µg/l hydroxocobalamin acetate, were recalcified and the thrombin generation was stopped after 10 min or 20 min by arginine as described under Methods. The approx. SC200 was 40 µg/l.

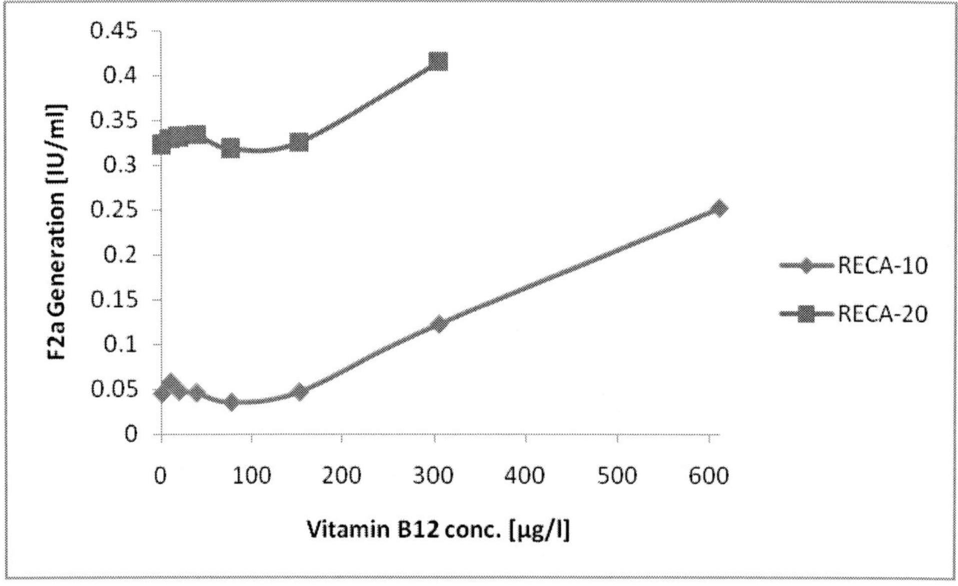

Figure 8. SC200 determination in plasma 7. 40 µl citrated plasma 7, supplemented with 0-610 µg/l cyanocobalamin, were recalcified and the thrombin generation was stopped after 10 min or 20 min by arginine as described under Methods. The approx. SC200 was 250 µg/l.

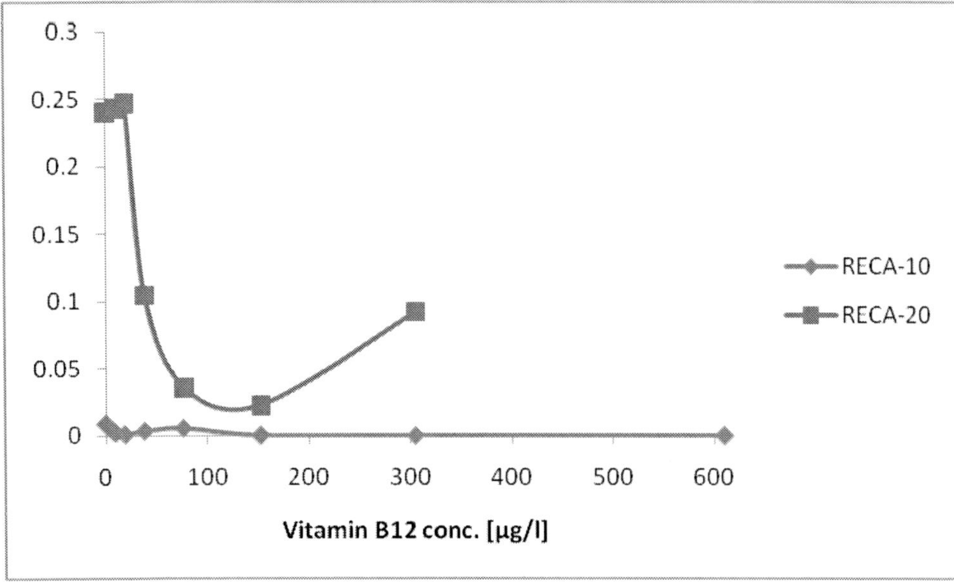

Figure 9. IC50, SC200 determination in plasma 8. 40 µl citrated plasma 8, supplemented with 0-610 µg/l cyanocobalamin, were recalcified and the thrombin generation was stopped after 10 min or 20 min by arginine as described under Methods. The approx. IC50 was 30 µg/l.

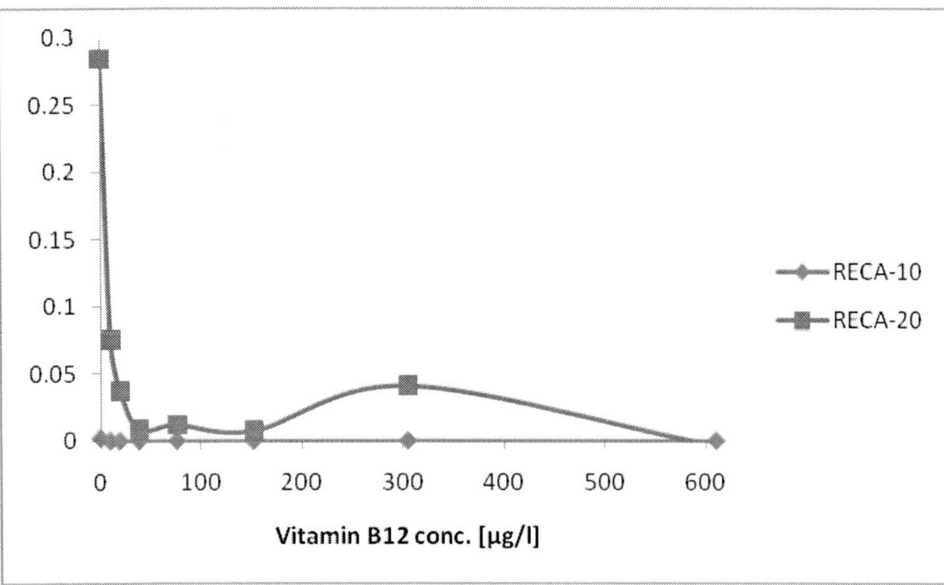

Figure 10. IC50, SC200 determination in plasma 9. 40 µl citrated plasma 9, supplemented with 0-610 µg/l cyanocobalamin, were recalcified and the thrombin generation was stopped after 10 min or 20 min by arginine as described under Methods. The approx. IC50 was 5 µg/l.

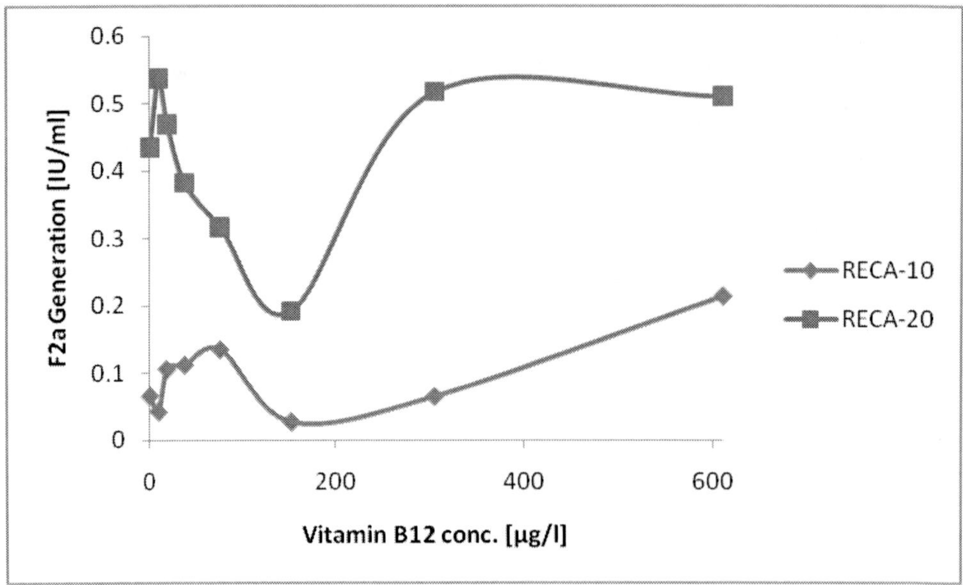

Figure 11. SC200, IC50 determination in plasma 10. 40 µl citrated plasma 10, supplemented with 0-610 µg/l cyanocobalamin, were recalcified and the thrombin generation was stopped after 10 min or 20 min by arginine as described under Methods. The approx. SC200 was 75 µg/l, the approx. IC50 was 130 µg/l.

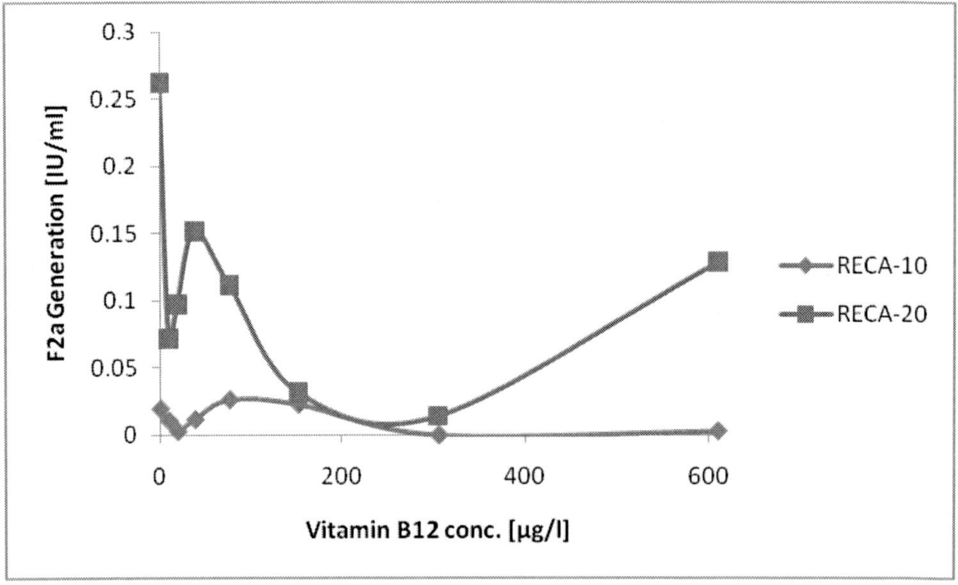

Figure 12. IC50 determination in plasma 11. 40 µl citrated plasma 11, supplemented with 0-610 µg/l cyanocobalamin, were recalcified and the thrombin generation was stopped after 10 min or 20 min by arginine as described under Methods. The approx. IC50 was 5 µg/l.

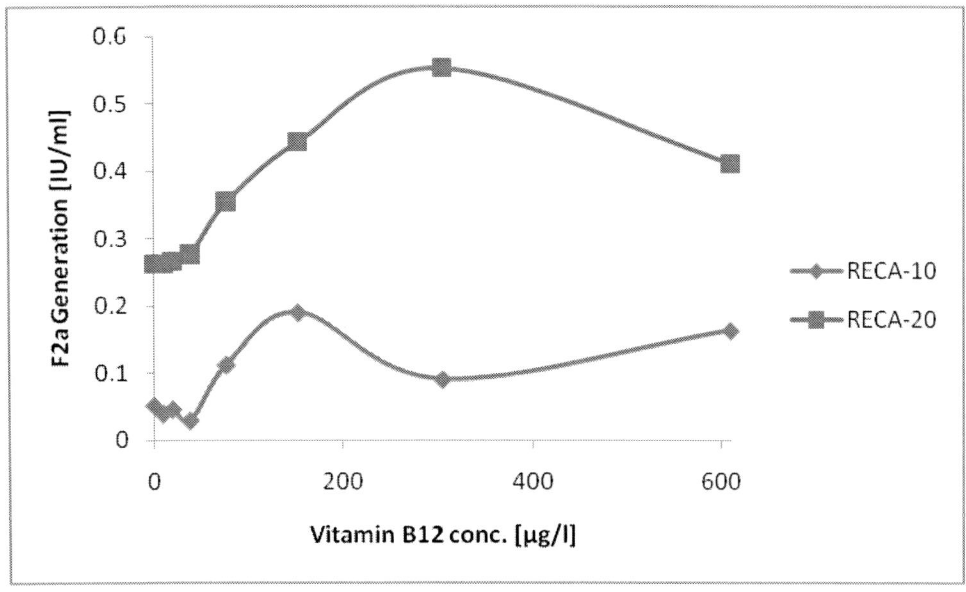

Figure 13. IC50 determination in plasma 12. 40 µl citrated plasma 12, supplemented with 0-610 µg/l cyanocobalamin, were recalcified and the thrombin generation was stopped after 10 min or 20 min by arginine as described under Methods. The approx. SC200 was 65 µg/l.

REFERENCES

[1]   www.wikipedia.org

[2]   Proinsias K, Giedyk M, Gryko D. Vitamin B$_{12}$: chemical modifications. *Chem Soc Rev.* 2013 May 29.

[3]   Jayaram N, Rao MG, Narasimha A, Raveendranathan D, Varambally S, Venkatasubramanian G, Gangadhar BN. Vitamin B$_{12}$ levels and psychiatric symptomatology: a case series. *J Neuropsychiatry Clin Neurosci.* 2013; 25: 150-2.

[4]   Glier MB, Green TJ, Devlin AM. Methyl nutrients, DNA methylation, and cardiovascular disease. *Mol Nutr Food Res.* 2013 May 10. doi: 10.1002/mnfr. 201200636.

[5]   www.AOK.de

[6]   Stief TW. Drug - induced thrombin generation: the breakthrough. *Hemostasis Laboratory* 2010; 3: 3-6.

[7]   Stief TW. Xenobiotic - induced pancreas carcinoma following mesenteric vein thrombosis : a hypothesis. *Hemostasis Laboratory* 2010; 3: 253-8.

[8]   Stief TW. Factor 12 activation in two purified systems. *Hemostasis Laboratory* 2011; 4: 385-408.

[9]   Stief TW. Zn^{2+}, hexane, or glucose activate factor 12 and/or prekallikrein in two purified systems. *Hemostasis Laboratory* 2011; 4: 409-26.

[10]   Stief TW. Zn^{2+}, hexane, valproate, or glucose in two purified systems of F12-PK-HMWK. *Hemostasis Laboratory* 2012; 5: 35-50.

[11] Stief TW, Mohrez M. HMWK increases of decreases F12a generation dependent on the contact trigger concentration in two purified systems. *Hemostasis Laboratory* 2012; 5: 51-65.

[12] Stief TW. Pathological thrombin generation by the synthetic inhibitor argatroban. *Hemostasis Laboratory* 2009; 2: 83-104.

[13] Stief TW, Brödje D. Thrombin generation by 3-hydroxy-butyrate, acetoacetate, or acetone. *Hemostasis Laboratory* 2013; 6: 17-50.

[14] Stief TW. Thrombin generation by creatinine. *Hemostasis Laboratory* 2011; 4: 191-9.

[15] Stief TW. Thrombin generation by therapeutic fibrinogen. *Hemostasis Laboratory* 2011; 4: 467-82.

[16] Stief TW. LMWH - action-monitoring for all patients. *Acta Paediatr.* 2012; 101: e314.

Chapter 10

DRAMATIC INCREASE OF BLOOD ROS GENERATION BY VITAMINS B_1, B_2, B_6, B_{12}

ABSTRACT

Background: The vitamins B_1, B_2, B_6, B_{12} act as coenzymes of important enzymes in neutrophil physiology. The up-regulation of the NADPH-oxidase, the physiologic generator system of large quantities of reactive oxygen species (ROS; H_2O_2 being the primary ROS and $^1\Delta O_2{}^*$ being the most selective ROS against "non-self") is the central activity of neutrophils upon contact with fungi, bacteria, or micro-thrombi. The regulation of $^1\Delta O_2{}^*$ generation by the vitamins B_1, B_2, B_6, B_{12} is the purpose of this work.

Material and Methods: 10 µl citrated blood (C1,C2,C3) or EDTA-blood were added to 120 µl modified (without phenol red) HBSS and 30 µl vitamin B_1, B_2, B_6, or B_{12} in 0.9% NaCl in black polystyrene plates (Brand®781608). After 0 min (BRGA) or 60 min (BRGA-60-) 10 µl 0.28 mM luminol (final conc.), 10 µl 2 µg/ml zymosan A (final conc.) were added. The luminescence was measured in the ascending part of the ROS generation kinetic by a photons-multiplying microtiter plate luminometer with an integration time of 0.5s per well and expressed in % control (unsupplemented blood).

Results: In the BRGA the mean approx. SC200 values of citrated blood were 12, 0.6, 4, 0.1 mg/l vitamin B_1, vitamin B_2, vitamin B_6, vitamin B_{12}. In the BRGA-60- the respective mean approx. SC200 values of citrated blood were 3, 0.17, 5, 0.1 mg/l. For EDTA-blood the values were 2, 0.1, 2, 0.05 versus 10, 0.5, 4, 0.1 mg/l. Seldomly, a plasma could react to a vitamin with an approx. IC50 instead of an approx. SC200.

Discussion: The vitamins studied react differently from individual to individual. Only by performing individual BRGA or BRGA-60- analysis the individual patient will know his personal vitamin dosage that stimulates or suppresses his neutrophils. This information is very important because without knowing the different individual reaction types a patient prone to hyper-inflammation could erroneously take a neutrophils-stimulating vitamin or a patient susceptible to hypo-inflammation could ingest a neutrophils suppressor.

Keywords: Vitamins B_1, B_2, B_6, B_{12}, reactive oxygen species, ROS, singlet oxygen, NADPH-oxidase, neutrophils

INTRODUCTION

The vitamins B_1, B_2, B_6, B_{12} in activated intracellular form are coenzymes of important enzymes in neutrophil physiology. The up-regulation of the NADPH-oxidase, the physiologic generator system of large quantities of reactive oxygen species (ROS; H_2O_2 being the primary ROS and $^1\Delta O_2^*$ being the most selective ROS against "non-self") is the central activity of neutrophils upon contact with fungi, bacteria, or micro-thrombi. The regulation of $^1\Delta O_2^*$ generation by the vitamins B_1, B_2, B_6, B_{12} [1-17] analyzed by the blood ROS generation assay (BRGA) or the BRGA with 60 min pre-incubation (BRGA-60-) is the purpose of this work.

MATERIAL AND METHODS

10 µl normal citrated blood (C1,C2,C3) or normal EDTA-blood (the blood samples were about 1d old) were added after written informed consent to 120 µl modified (without phenol red) Hanks′ Balanced Salt Solution (SAFC Biosciences – Sigma, Deisenhofen, Germany; article nr. 55037C-1000ML; HBSS) and 30 µl vitamin B_1 (thiamin; ratiopharm, Ulm, Germany; 2 ml ampoules of 50 mg/ml) (final conc. 0-104 mg/l), vitamin B_2 (riboflavin; Streuli, Uznach, Switzerland; 10 mg amount in 2 ml ampoules) (final conc. 0-10.4 mg/l), vitamin B_6 (pyridoxine hydrochloride; ratiopharm, Ulm, Germany; 2 ml ampoules of 50 mg/ml) (final conc. 0-104 mg/l), or vitamin B_{12} (hydroxocobalamin acetate from Pascoe, Giessen, Germany (1 ml ampoules of 1500 µg) or cyanocobalamin from Jenapharm, Brehna, Germany; (1 ml ampoules of 1000 µg)) (final conc. 0-2.1 mg/l) in black high quality polystyrene plates (Brand, Wertheim, Germany; article nr. 781608). After 0 min (BRGA) or 60 min (BRGA-60-) 10 µl 0.28 mM (final conc.) luminol (Sigma), 10 µl 2 µg/ml (final conc.) zymosan A (Sigma) were added. The luminescence was measured in the ascending part of the ROS generation kinetic by a photons-multiplying microtiter plate luminometer (LUmo; Autobio-anthos; Krefeld, Germany) with an integration time of 0.5s per well and expressed in % control (unsupplemented blood) [17]. The approximate 200% stimulatory or 50% inhibitory concentrations of the serine proteases on blood ROS generation were determined (approx. SC200, approx. IC50).

RESULTS AND DISCUSSION

Figures 1-4 and Table 1 demonstrate the stimulation of ROS generation measured by the BRGA. The approx. SC200 values for vitamin B_1 were 5 mg/l, 10 mg/l, 20 mg/l in citrated blood, 2 mg/l in EDTA-blood (Figure 1).

The approx. SC200 values for vitamin B_2 were 0.3 mg/l, 1.2 mg/l, 0.3 mg/l in citrated blood, 0.1 mg/l in EDTA-blood. For C2 there appeared also an approx. IC50 of 0.4 mg/l (Figure 2).

The approx. SC200 values for vitamin B_6 were 1.5 mg/l, - mg/l, 6 mg/l in citrated blood, 2 mg/l in EDTA-blood. For C2 there appeared an approx. IC50 of 55 mg/l (Figure 3).

The approx. SC200 values for vitamin B_{12} were 0.1 mg/l, - , 0.1 mg/l in citrated blood, 0.05 mg/l in EDTA-blood. For C2 there appeared an approx. IC50 of 0.1 mg/l. Both vitamin B_{12} forms behaved similarly (Figure 4).

Table 1. Approx. SC200 values of different B vitamins on blood singlet oxygen generation in BRGA

Vitamin	Approx. SC200 citrated blood [mg/l]	Approx. SC200 EDTA-blood [mg/l]
B_1	12±8	2
B_2	0.6±0.5	0.1
B_6	4±3	2
Cyanocobalamin	0.1	0.05
Hydroxocobalamin	0.1	0.05

Table 2. Approx. SC200 values of different B vitamins on blood singlet oxygen generation in BRGA-60-

Vitamin	Approx. SC200 citrated blood	Approx. SC200 EDTA-blood
B_1	3±1	10
B_2	0.17±0.06	0.5
B_6	5±7	4
Cyanocobalamin	0.1±0.1	0.1
Hydroxocobalamin	0.1±0.1	0.1

In conclusion, usually the B vitamins here studied suppressed blood ROS generation, however with great inter-individual variations; sometimes there occurred even an inhibitory action of the vitamin instead of a stimulation. The individual blood has to be individually studied to know the personal medical situation of the patient. This information is very important because without knowing the different individual reaction types a patient prone to hyper-inflammation/coagulation activation could erroneously take a neutrophils-stimulating coagulation-triggering vitamin [18-21] or a patient susceptible to hypo-inflammation could ingest a neutrophils-suppressor.

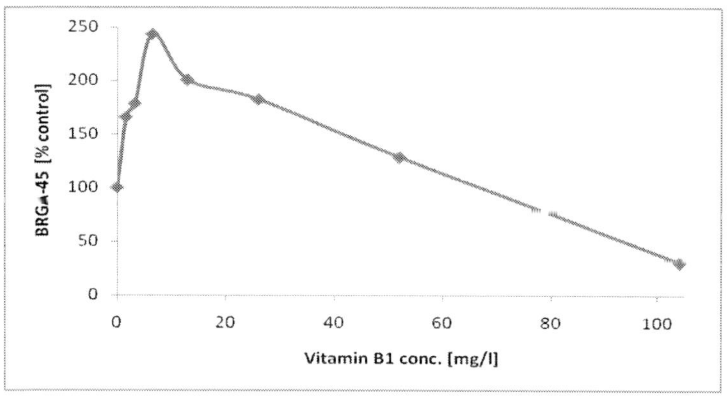

Figure 1. (Continued)

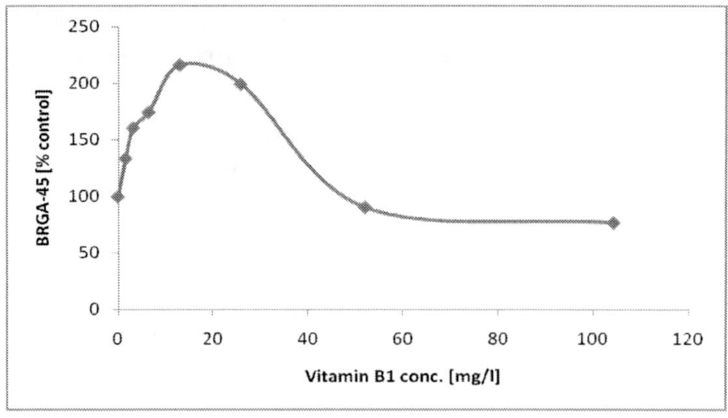

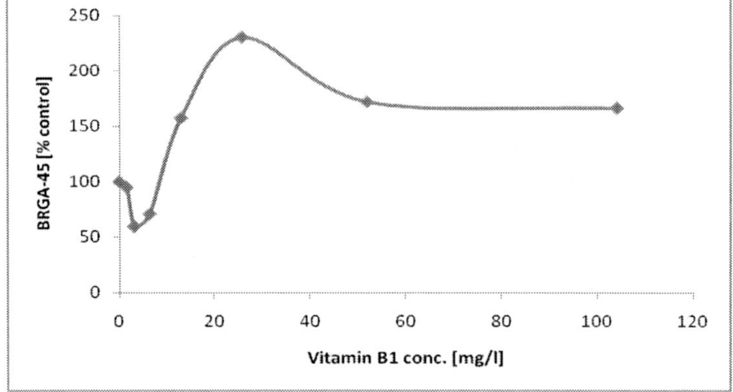

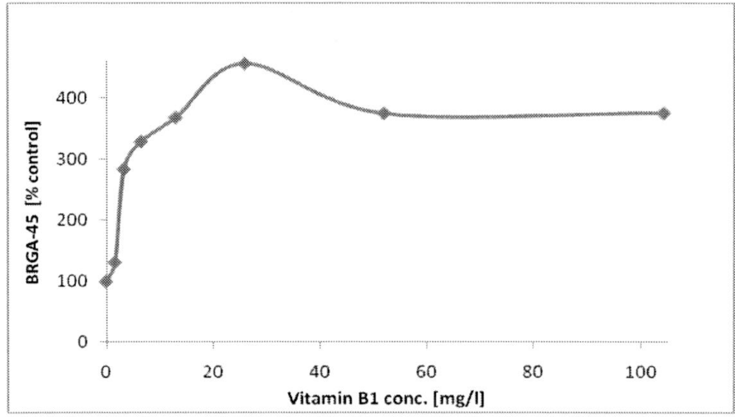

Figure 1. Stimulation of blood ROS generation by vitamin B1. 10 µl citrated blood C1, C2, C3 or EDTA-blood E1 were added to 0-104 mg/l vitamin B1 in HBSS. 10 µl 0.28 mM luminol and 10 µl 2 µg/ml zymosan A were added. At 45 min (37°C) incubation time of the blood ROS generation assay (BRGA-45) the luminescences were measured by photons-enhancing microtiter plate luminometer. The light emissions were expressed in % control = unsupplemented blood (the 100% control values were for C1,C2 about 1500 RLU/s for C3, E1 about 700 RLU/s at BRGA-45; at BRGA-97 (maximum) about 5600 RLU/s (C1,C2,C3), about 1300 RLU/s (E1). The approx. SC200 values were 5 mg/l, 10 mg/l, 20 mg/l in citrated blood (Figures 1a-c), 2 mg/l in EDTA-blood (Figure 1d).

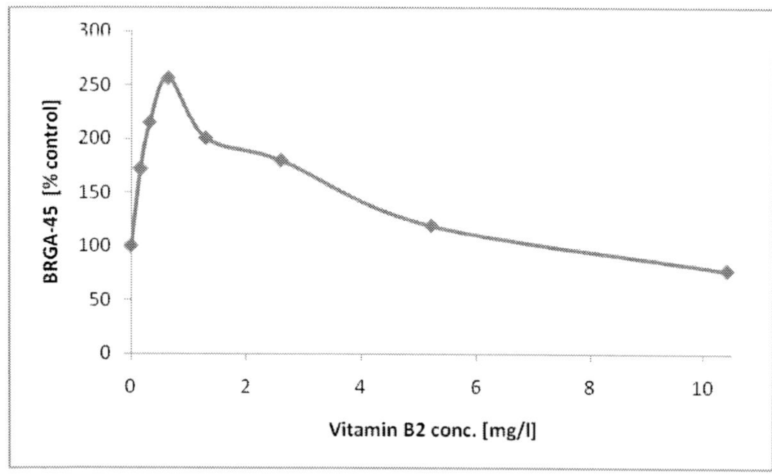

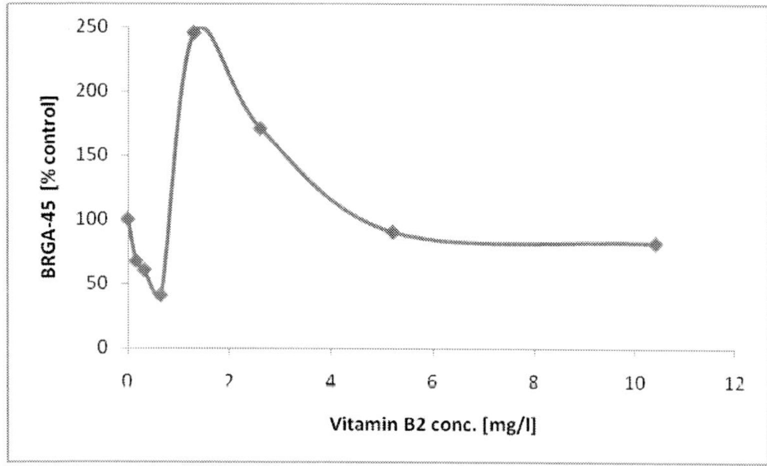

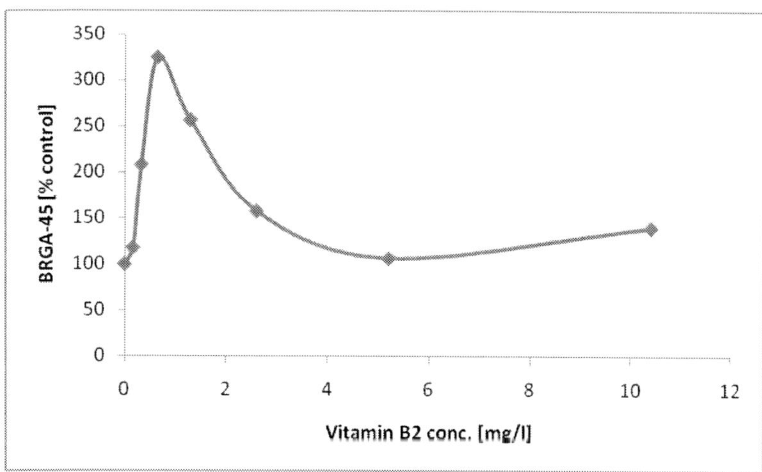

Figure 2. (Continued)

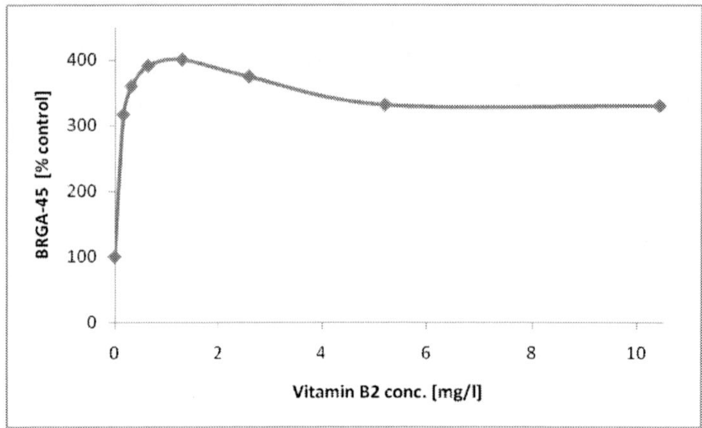

Figure 2. Modulation of blood ROS generation by vitamin B_2. 10 μl citrated blood C1, C2, C3 or EDTA-blood E1 were added to 0-10.4 mg/l vitamin B_2 in HBSS. 10 μl 0.28 mM luminol and 10 μl 2 μg/ml zymosan A were added. At 45 min (37°C) incubation time of the blood ROS generation assay (BRGA-45) the luminescences were measured by photons-enhancing microtiter plate luminometer. The approx. SC200 values were 0.3 mg/l, 1.2 mg/l, 0.3 mg/l in citrated blood (Figures 2a-c), 0.1 mg/l in EDTA-blood (Figure d). For C2 there appeared also an approx. IC50 of 0.4 mg/l.

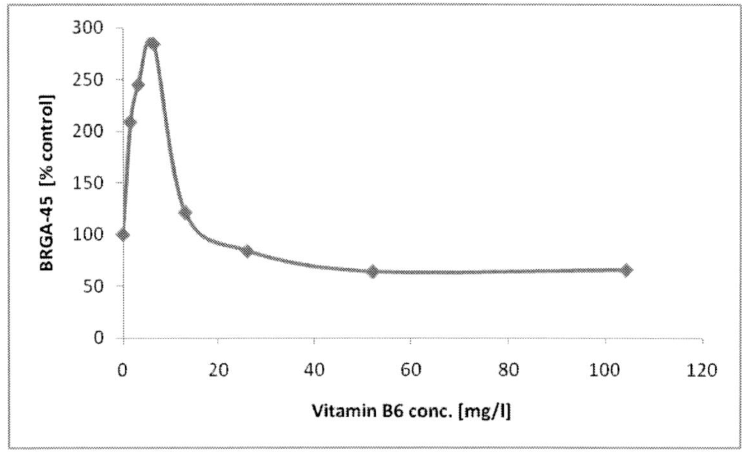

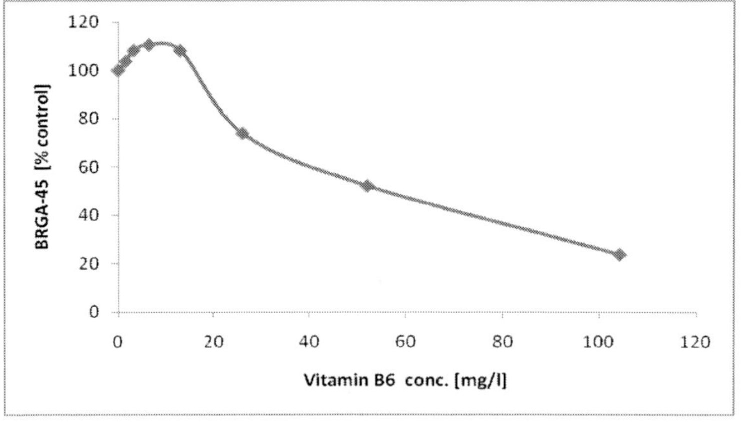

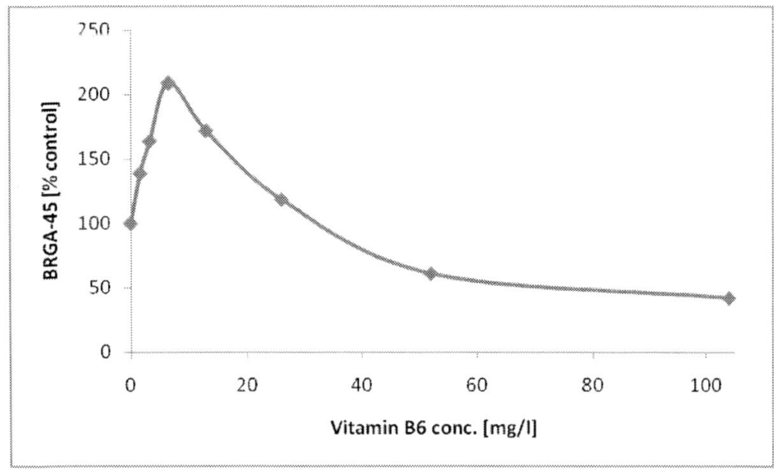

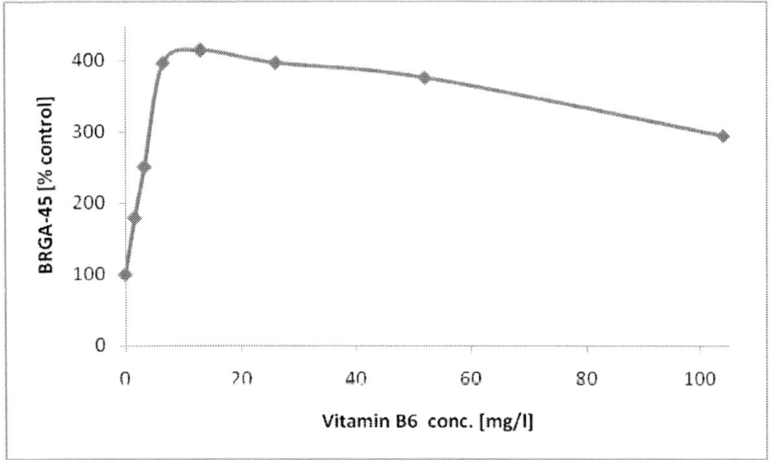

Figure 3. Modulation of blood ROS generation by vitamin B_6. 10 µl citrated blood C1, C2, C3 or EDTA-blood E1 were added to 0-104 mg/l vitamin B_6 in HBSS. 10 µl 0.28 mM luminol and 10 µl 2 µg/ml zymosan A were added. At 45 min (37°C) incubation time of the blood ROS generation assay (BRGA-45) the luminescences were measured by photons-enhancing microtiter plate luminometer. The approx. SC200 values were 1.5 mg/l, - mg/l, 6 mg/l in citrated blood (Figures 3a-c), 2 mg/l in EDTA-blood (Figure 3a). For C2 there appeared an approx. IC50 of 55 mg/l.

Figures 5-8 and Table 2 demonstrate the stimulation of ROS generation measured by the BRGA. The approx. SC200 values for vitamin B_1 were 2 mg/l, 4 mg/l, 2 mg/l in citrated blood, 10 mg/l in EDTA-blood (Figure 5).

The approx. SC200 values for vitamin B_2 were 0.2 mg/l, 0.1 mg/l, 0.2 mg/l in citrated blood, 0.5 mg/l in EDTA-blood (Figure 6). The approx. SC200 values for vitamin B_6 were 2 mg/l, 1 mg/l, 13 mg/l in citrated blood, 4 mg/l in EDTA-blood (Figure 7).

The approx. SC200 values were for hydroxocobalamin 0.03 mg/l, 0.02, 0.2 mg/l in citrated blood, 0.1 mg/l in EDTA-blood. For cyanocobalamin the approx. SC200 values were 0.03 mg/l, - , 0.2 mg/l in citrated blood, 0.1 mg/l in EDTA-blood. For C2 an approx. IC50 of 0.3 mg/l appeared, for EDTA-blood there was an approx. IC50 of 0.7 mg/l (Figure 8).

The vitamins B_1 and B_2 were activated during the pre-incubation phase of BRGA-60-. Their approx. SC200 decreased about 4fold (Tables 1, 2).

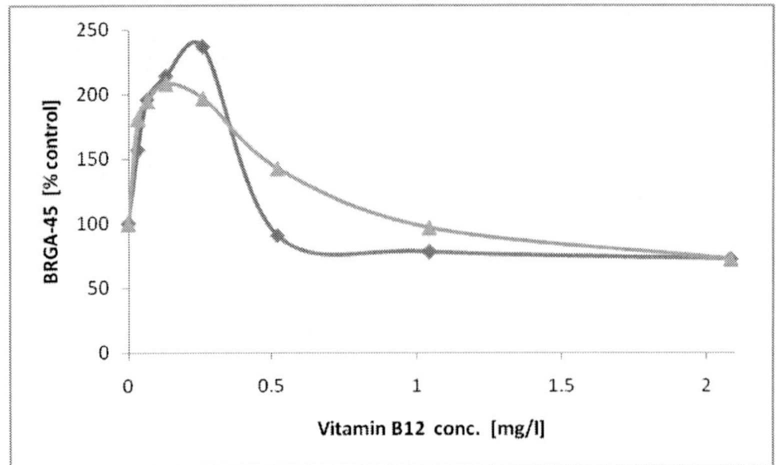

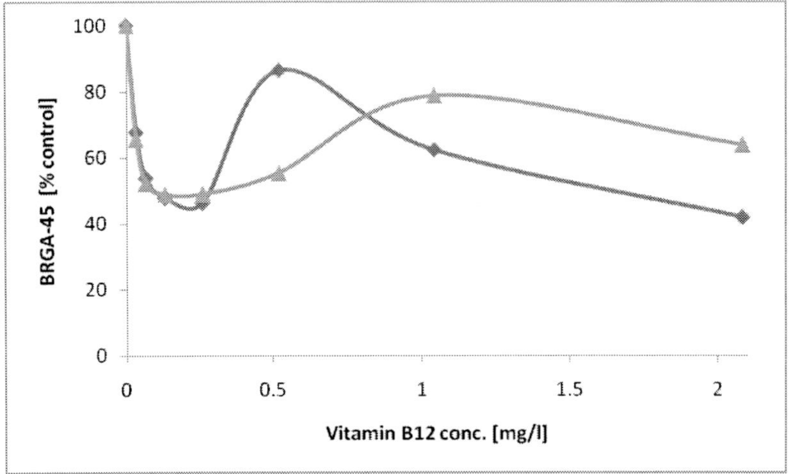

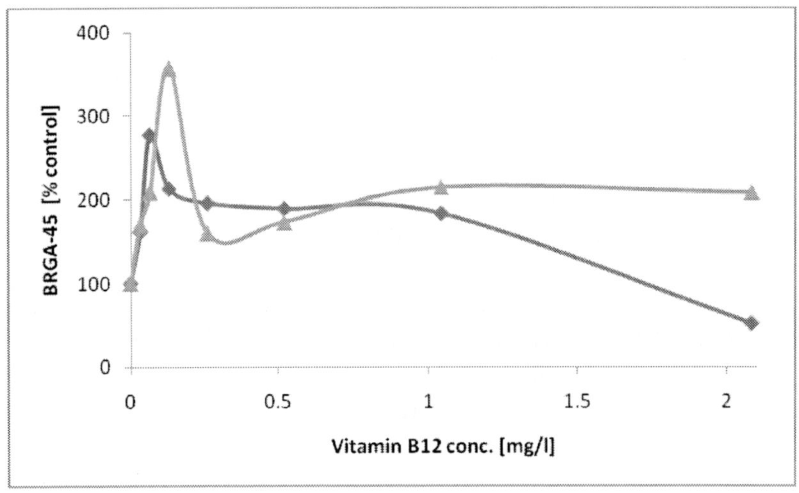

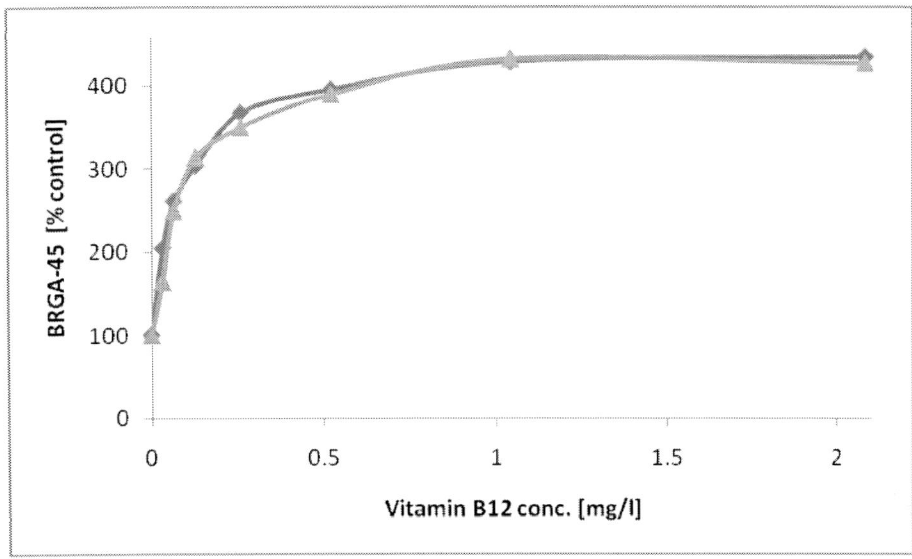

Figure 4. Stimulation of blood ROS generation by vitamin B$_{12}$. 10 µl citrated blood C1, C2, C3 or EDTA-blood E1 were added to 0-2.1 mg/l vitamin B$_{12}$ (hydroxocobalamin ♦, cyanocobalamin ▲) in HBSS. 10 µl 0.28 mM luminol and 10 µl 2 µg/ml zymosan A were added. At 45 min (37°C) incubation time of the blood ROS generation assay (BRGA-45) the luminescences were measured by photons-enhancing microtiter plate luminometer. The approx. SC200 values were 0.1 mg/l, - , 0.1 mg/l in citrated blood (Figures 4a-c), 0.05 mg/l in EDTA-blood (Figure 4d). For C2 there appeared an approx. IC50 of 0.1 mg/l. Both vitamin B$_{12}$ forms behaved similarly.

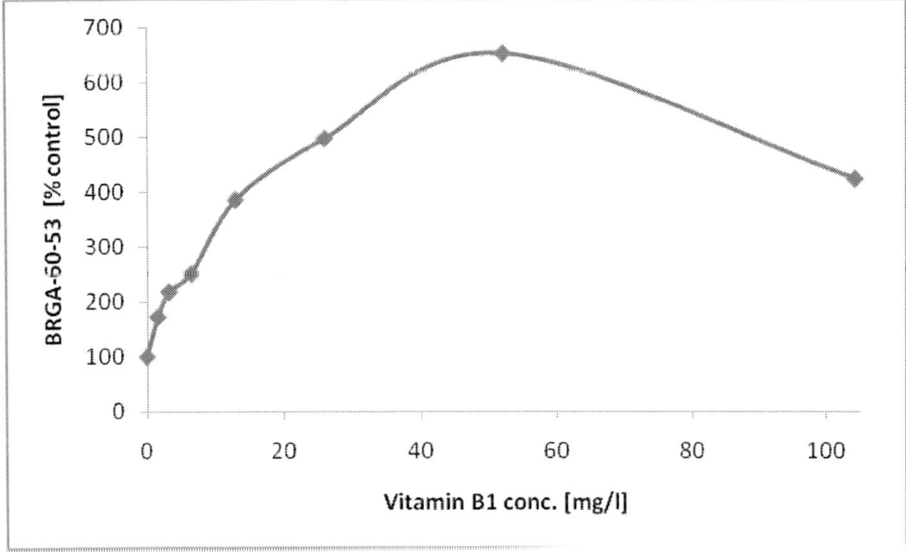

Figure 5. (Continued)

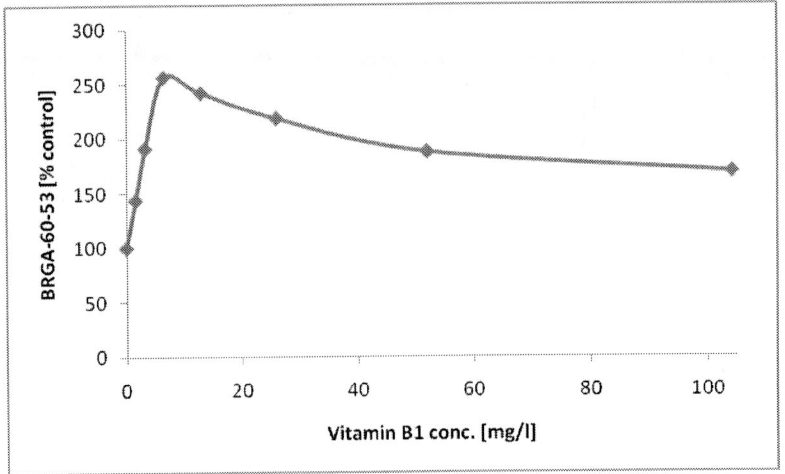
300
250
200
150
100
50
0
BRGA-60-53 [% control]
0   20   40   60   80   100
Vitamin B1 conc. [mg/l]

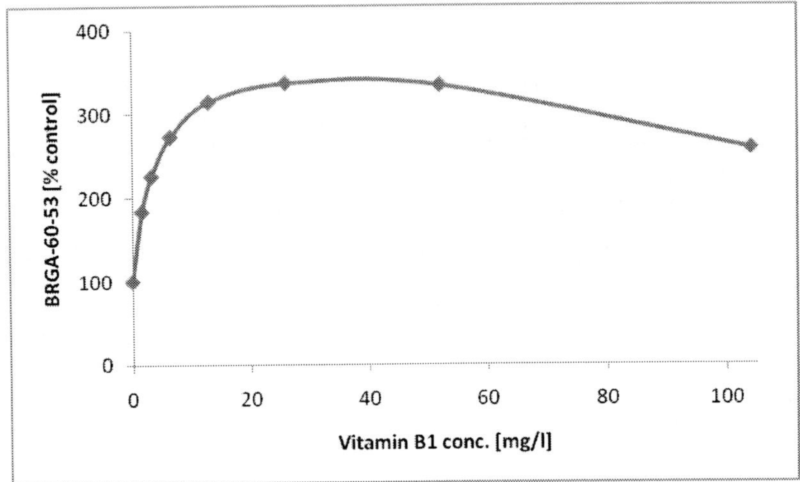
400
300
200
100
0
BRGA-60-53 [% control]
0   20   40   60   80   100
Vitamin B1 conc. [mg/l]

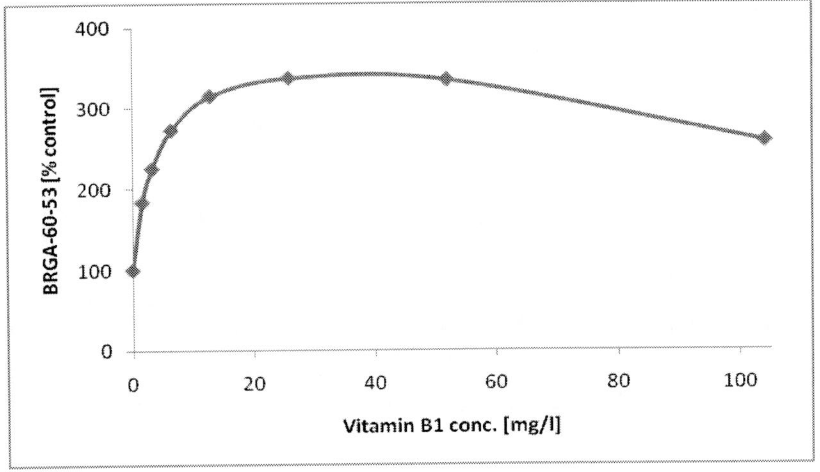
400
300
200
100
0
BRGA-60-53 [% control]
0   20   40   60   80   100
Vitamin B1 conc. [mg/l]

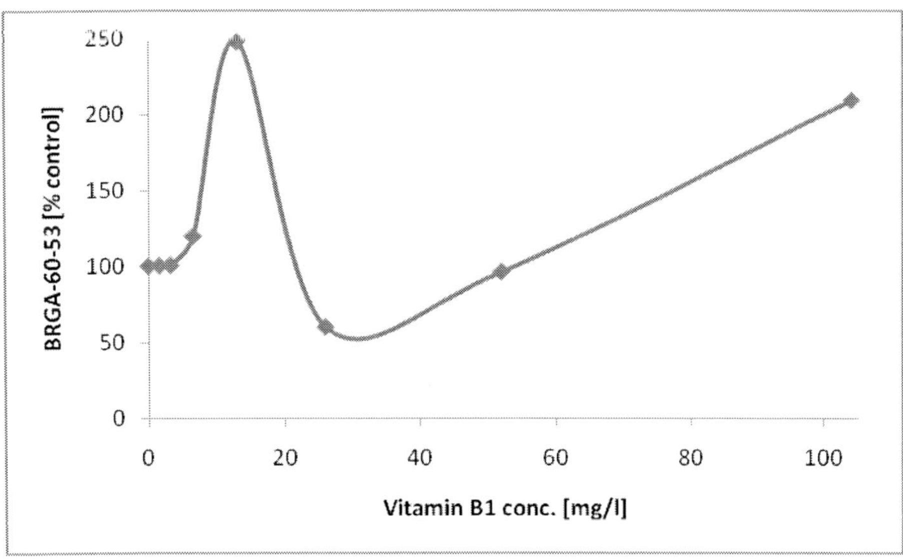

Figure 5. Stimulation of pre-incubated blood ROS generation by vitamin B_1. 10 µl citrated blood C1, C2, C3 or EDTA-blood E1 were added to 0-104 mg/l vitamin B_1 in HBSS. After 60 min (37°C) 10 µl 0.28 mM luminol and 10 µl 2 µg/ml zymosan A were added. At 53 min (37°C) incubation time of the blood ROS generation assay (BRGA-60-53) the luminescences were measured by photons-enhancing microtiter plate luminometer. The light emissions were expressed in % control = unsupplemented blood; the 100% control values / maxima at BRGA-75 were about 1100/2100 RLU/S (C1), about 700/1700 RLU/s (C2); about 2100/2700 RLU/s (C3); about 1500/1900 RLU/s (E1). The approx. SC200 values were 2 mg/l, 4 mg/l, 2 mg/l in citrated blood (Figures 5a-c), 10 mg/l in EDTA-blood (Figure 5d).

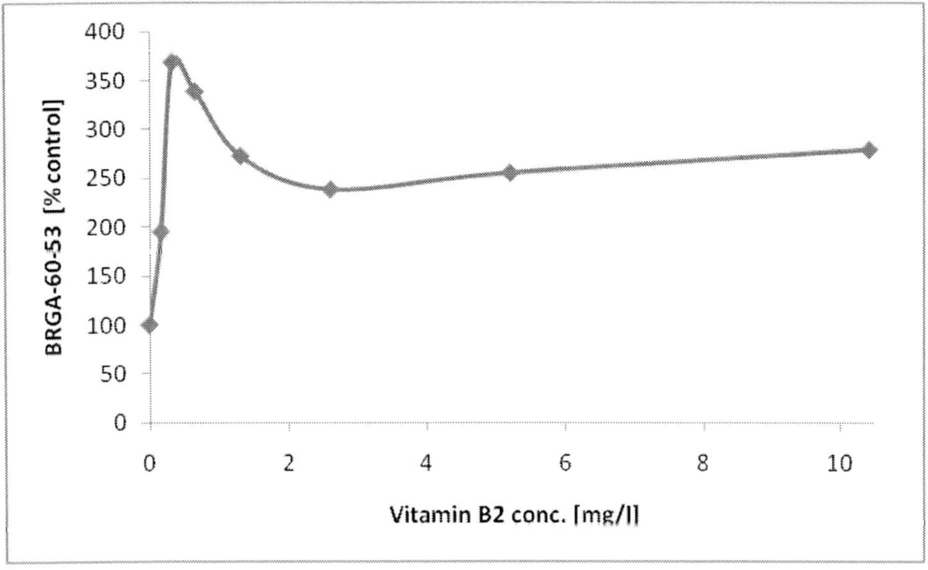

Figure 6. (Continued)

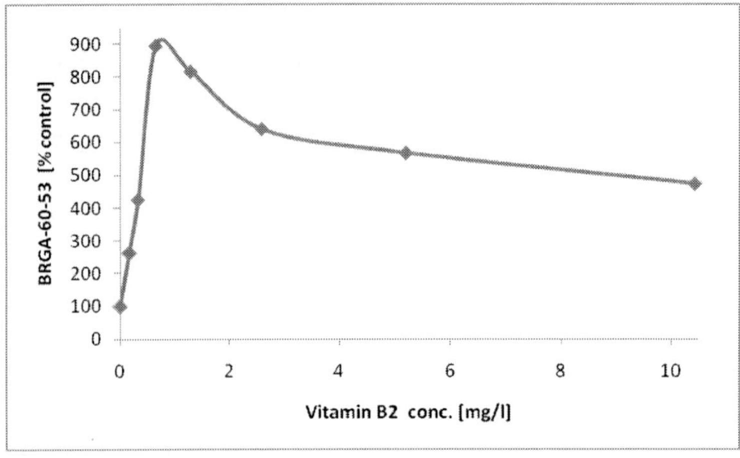

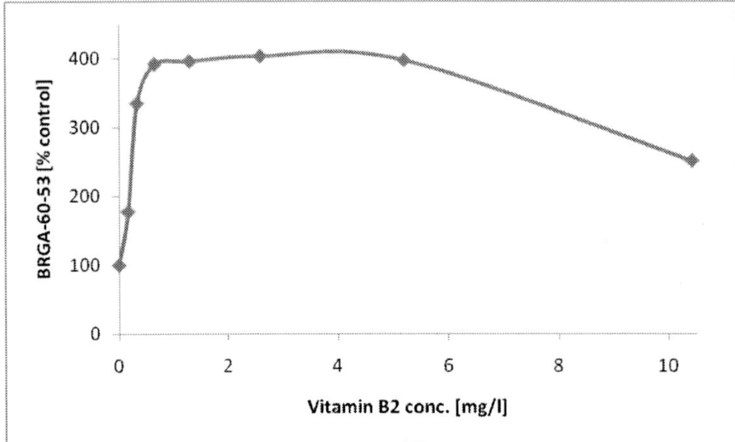

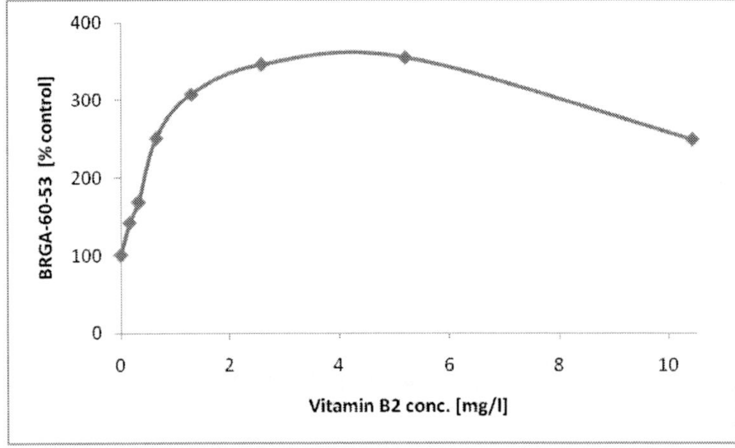

Figure 6. Stimulation of pre-incubated blood ROS generation by vitamin B_2. 10 µl blood C1, C2, C3, or E1 were added to 0-10.4 mg/l vitamin B_2 in HBSS. After 60 min (37°C) 10 µl 0.28 mM luminol and 10 µl 2 µg/ml zymosan A were added. At 53 min (37°C) incubation time of the blood ROS generation assay (BRGA-60-53) the luminescences were measured by photons-enhancing microtiter plate luminometer. The approx. SC200 values were 0.2 mg/l, 0.1 mg/l, 0.2 mg/l in citrated blood (Figures 6a-c), 0.5 mg/l in EDTA-blood (Figure 6d).

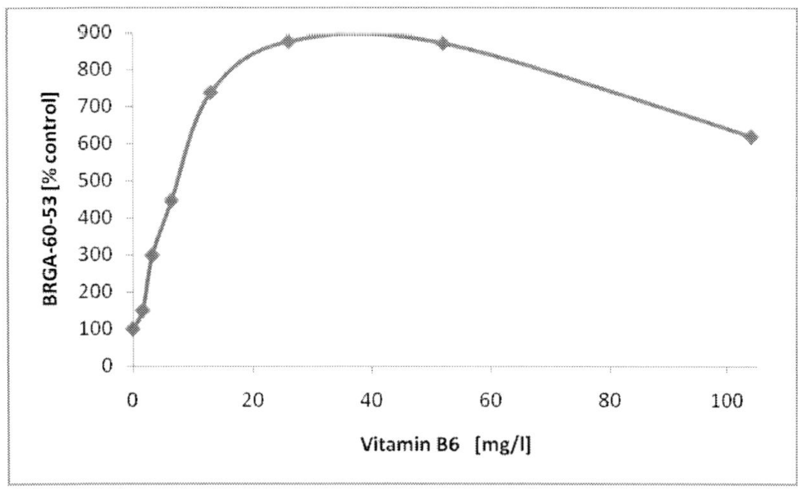

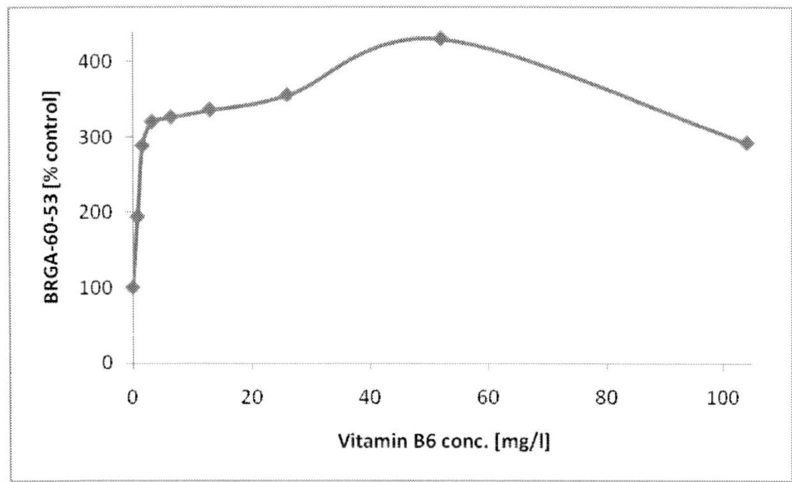

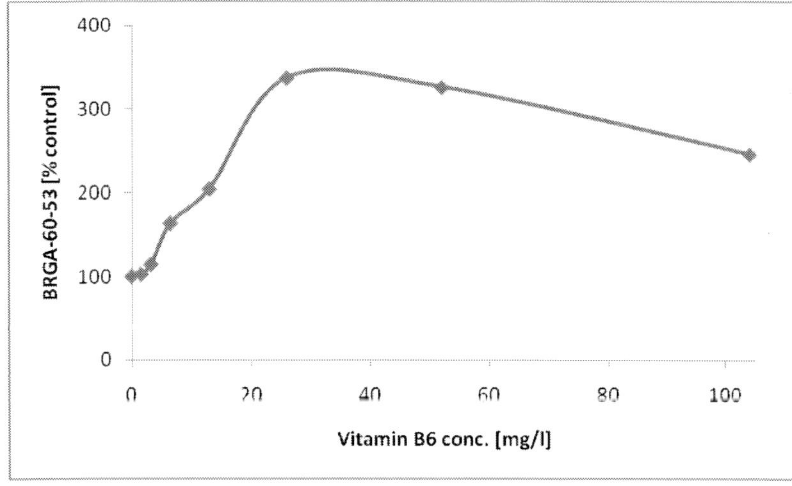

Figure 7. (Continued).

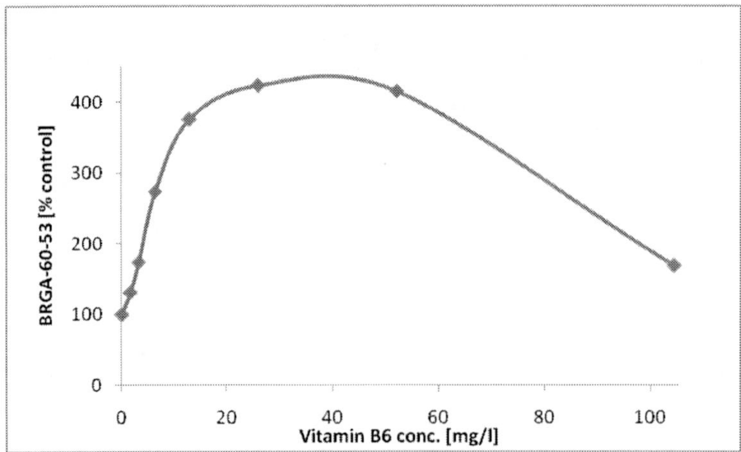

Figure 7. Stimulation of pre-incubated blood ROS generation by vitamin B_6. 10 µl blood C1, C2, C3, or E1 were added to 0-104 mg/l vitamin B_2 in HBSS. After 60 min (37°C) 10 µl 0.28 mM luminol and 10 µl 2 µg/ml zymosan A were added. At 53 min (37°C) incubation time of the blood ROS generation assay (BRGA-60-53) the luminescences were measured by photons-enhancing microtiter plate luminometer. The approx. SC200 values were 2 mg/l, 1 mg/l, 13 mg/l in citrated blood (Figures 7a-c), 4 mg/l in EDTA-blood (Figure 7d).

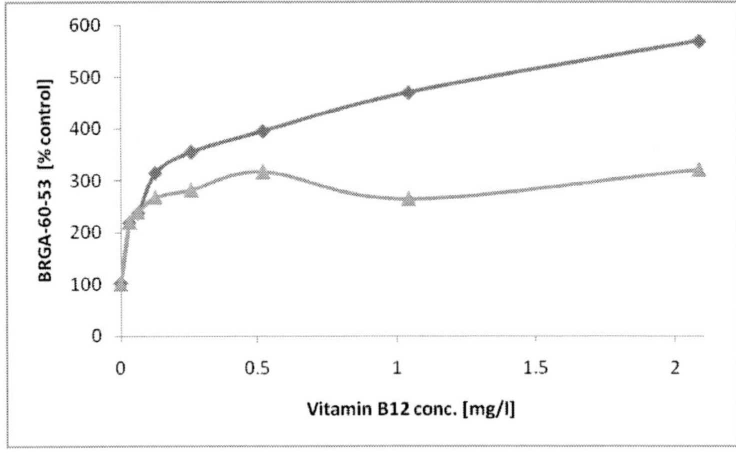

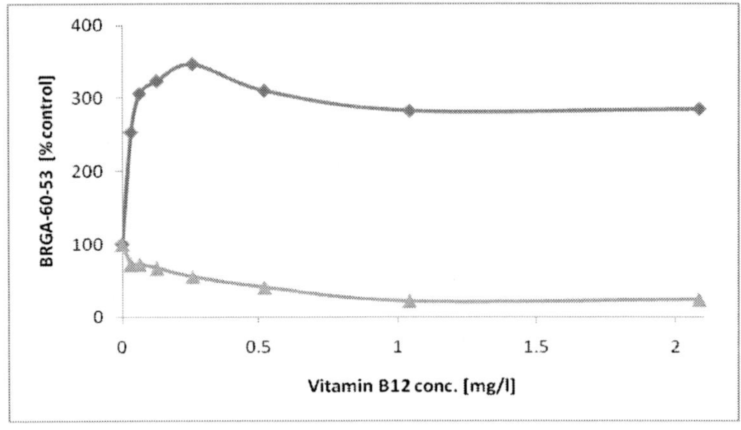

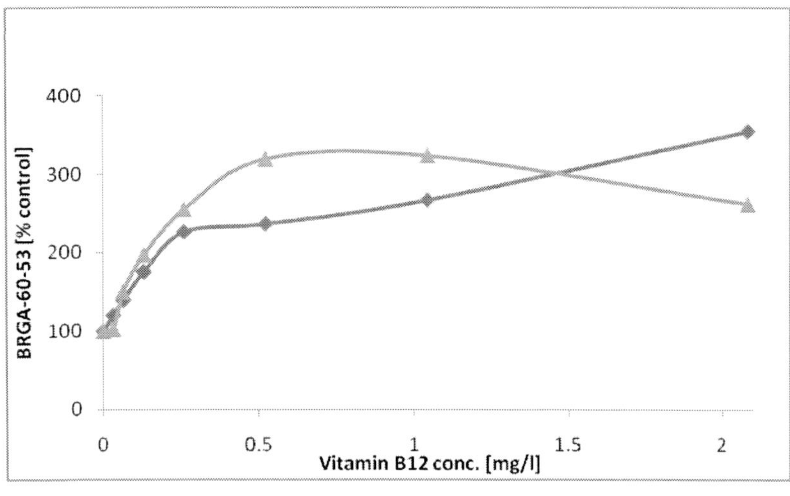

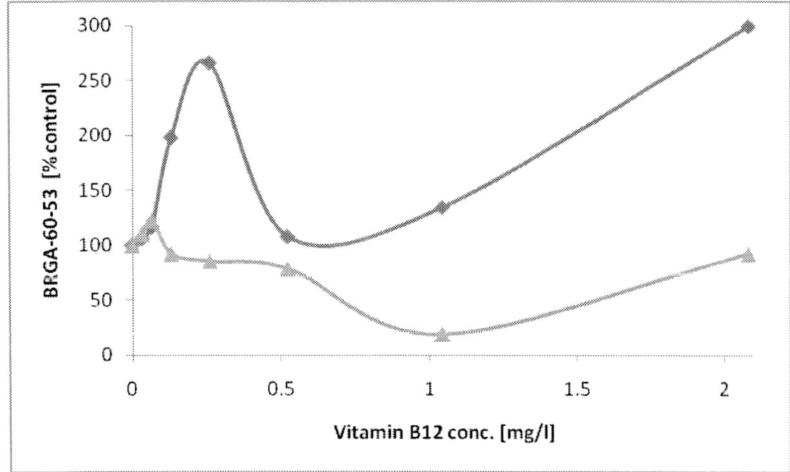

Figure 8. Stimulation of pre-incubated blood ROS generation by vitamin B_{12}. 10 µl citrated blood C1, C2, C3 or EDTA-blood E1 were added to 0-2.1 mg/l vitamin B_{12} (hydroxocobalamin ♦, cyanocobalamin ▲) in HBSS. After 60 min 10 µl 0.28 mM luminol and 10 µl 2 µg/ml zymosan A were added. At 53 min (37°C) incubation time of the blood ROS generation assay (BRGA-60-53) the luminescences were measured by photons-enhancing microtiter plate luminometer. The approx. SC200 values were for hydroxocobalamin 0.03 mg/l, 0.02, 0.2 mg/l in citrated blood (Figures 8a-c), 0.1 mg/l in EDTA-blood (Figure 8d). For cyanocobalamin the approx. SC200 values were 0.03 mg/l, - , 0.2 mg/l in citrated blood, 0.1 mg/l in EDTA-blood. For C2 an approx. IC50 of 0.3 mg/l appeared, for EDTA-blood there was an approx. IC50 of 0.7 mg/l.

REFERENCES

[1]   Stief TW. Kallikrein activates prothrombin. *Clin Appl Thrombosis/Hemostasis.* 2008: 14: 97-8.

[2]   Stief TW. Kallikrein triggers thrombin generation. *Hemostasis Laboratory* 2009; 2: 45-56.

[3] Stief TW. Zn^{2+}, hexane, or glucose activate factor 12 and/or prekallikrein in two purified systems. *Hemostasis Laboratory* 2011; 4: 409-26.

[4] Stief TW. Kallikrein activates factor 10. *Hemostasis Laboratory* 2012; 5: 211-8.

[5] Stief TW, Klingmüller V. Diagnostic ultrasound activates pure prekallikrein. *Blood Coagulation and Fibrinolysis* 2012; 23: 781-3.

[6] Stief T. Hemostasis activation in inflammation: kallikrein, thrombin, $^{1}\Delta O_2^{*}$/hv. In: *Thrombin and Singlet Oxygen ($^{1}\Delta O_2^{*}$) Main Factors of Hemostasis*. Stief T, ed.; Nova Science Publishers, New York, 2013.

[7] Stief TW. Coumarins trigger altered matrix (AM) coagulation activation. *Hemost Lab.* 2013; 6: 121-8.

[8] Wachtfogel YT, Kettner C, Hack CE, Nuijens JH, Reilly TM, Knabb RM, Kucich U, Niewiarowski S, Edmunds LH Jr, Colman RW. Thrombin and human plasma kallikrein inhibition during simulated extracorporeal circulation block platelet and neutrophil activation. *Thromb Haemost.* 1998; 80: 686-91.

[9] Stief TW. Neutrophil granulocytes in hemostasis. *Hemostasis Laboratory* 2008; 1: 269-89.

[10] Nathan C. Neutrophils and immunity: challenges and opportunities. *Nat Rev Immunol.* 2006; 6: 173-82.

[11] Stief TW. The blood fibrinolysis / deep-sea analogy: a hypothesis on the cell signals singlet oxygen/photons as natural antithrombotics. *Thromb Res* 2000; 99: 1-20.

[12] Stief TW, Fareed J. The antithrombotic factor singlet oxygen/light ($^{1}O_2$/hv). *Clin Appl Thrombosis/Hemostasis* 2000; 6: 22-30.

[13] Stief TW, Fu K, Doss MO, Fareed J. The anti-thrombotic factor singlet oxygen ($^{1}O_2$) induces selective thrombolysis *in vivo* by massive phagocyte infiltration into the thrombus. *XVII Congress of the International Society on Thrombosis and Haemostasis*; Washington; August 14-21, 1999.

[14] Stief TW. The physiology and pharmacology of singlet oxygen. *Med Hypoth* 2003; 60: 567-572.

[15] Stief TW. Regulation of hemostasis by singlet oxygen ($^{1}\Delta O_2$). *Curr Vasc Pharmacol* 2004; 2: 357-362.

[16] Stief T. Micro-thrombi stimulate blood ROS generation. *Hemostasis Laboratory* 2013; 6: 315-25.

[17] Stief T. The routine blood ROS generation assay (BRGA) triggered by typical septic concentrations of zymosan A. *Hemostasis Laboratory* 2013; 6: 89-98.

[18] Stief T. Vitamin B_1 (Thiamine) Triggers AM-Coagulation. *Hemost Lab.* 2013; 7 (issue 3).

[19] Stief T. Vitamin B_2 (Riboflavin) Triggers AM-Coagulation. *Hemost Lab.* 2013; 7 (issue 3).

[20] Stief T. Vitamin B_6 Triggers AM-Coagulation. *Hemost Lab.* 2013; 7 (issue 3).

[21] Stief T. Vitamin B_{12} Triggers AM-Coagulation. *Hemost Lab.* 2013; 7 (issue 3).

Chapter 11

DRINKING WATER OR BEVERAGE ANALYSIS FOR INTRINSIC COAGULATION ACTIVATION

ABSTRACT

Background: Alteration of blood matrix is an important trigger of the intrinsic hemostasis cascade. The ingestion of large quantities of beverages changes blood matrix. Therefore, a routine test was established to analyze the action of plasma dilution by different drinks.

Material and Methods: 40 µl pooled normal citrated plasma or 40 mIU/ml bovine thrombin in 5% human albumin standard were supplemented with 0-40 µl a) 0.9% NaCl, b) H_2O, c) drinking water from a german village, d) drinking water from a german city, e) Pepsi Cola light, or f) Pepsi Cola in high quality U-wells polystyrene plates (Brand®781600). 4 µl 250 mM $CaCl_2$ in siliconized glass bottles were added. After 0 or 12 min 80 µl 2.5 M arginine, 0.16% Triton X 100, pH 8.6 were added. After 3 min 20 µl 1 mM HD-CHG-Ala-Arg-pNA in 1.25 M arginine, pH 8.7 were added and $\Delta A_{405\,nm}$ was determined.

Results and Discussion: 0.9% NaCl had the best action on thrombin generation. Addition of 5 µl volume to 40 µl plasma did not change thrombin generation. Drinking water 1 was much better than drinking water 2; at 2 µl supplementation DW1 had 118% thrombin generation, whereas DW2 had 188% thrombin generation. This thrombin activity was comparable to that induced by Pepsi Cola light (assuming that the beverage is intestinally totally absorbed). In conclusion, the present easy technique allows to analyze the quality of drinking water or beverages respective thrombin generation of blood plasma. Solutions similar to 0.9% NaCl are of clinical interest in (bleeding-unrelated) hypovolemic shock.

INTRODUCTION

Contact phase coagulation is very sensible to changes of blood matrix. Drinking large quantities of any beverage results in plasma dilution, diluted plasma facilitates the folding of factor 12 into F12a or of pre-kallikrein into kallikrein, the two starter enzymes of intrinsic hemostasis. Drinks may harm the organism by inducing a pro-thrombotic state.

Therefore, a routine test has been established and is now further investigated that analyzes the action of plasma dilution by different beverages [1].

MATERIAL AND METHODS

40 µl pooled normal citrated plasma (-30°C frozen/23°C thawed) of 12 healthy donors that gave written informed consent or 40 mIU/ml bovine thrombin (Siemens Healthcare, Munich, Germany) in 5% human albumin (CSL Behring, Marburg, Germany) standard were supplemented with 0-30 µl a) 0.9% NaCl (Braun, Melsungen, Germany), b) H_2O (Braun), c) drinking water from the village Pohlheim-Grüningen (in a commercial glass bottle), d) drinking water from the city of Marburg (in a commercial glass bottle), e) Pepsi Cola light (0.5 l plastic bottle), or f) Pepsi Cola (0.5 l plastic bottle) in high quality U-wells polystyrene plates (Brand, Wertheim, Germany; article nr. 781600). 4 µl 250 mM $CaCl_2$ in siliconized glass bottles (Siemens Healthcare) were added. After 0 or 12 min 80 µl 2.5 M arginine, 0.16% Triton X 100, pH 8.6 (Sigma, Deisenhofen, Germany) were added. After 3 min at room temperature 20 µl 1 mM fast chromogenic thrombin substrate HD-CHG-Ala-Arg-pNA (Pentapharm, Basel, Switzerland) in 1.25 M arginine, pH 8.7 were added and $\Delta A_{405\ nm}$ was determined by a microtiter plate photometer with a 1 mA resolution (PHOmo; Autobio-anthos, Krefeld, Germany). Considered were only thrombin generations in the ascending part of the coagulation reaction time versus thrombin activity curve [2,3].

RESULTS AND DISCUSSION

0.9% NaCl had the best action on thrombin generation (Figures 1,2). Addition of 5 µl volume to 40 µl plasma did not change thrombin generation. This relatively inert behavior of 0.9% NaCl has already been seen in plasma and in the purified system [4-6].

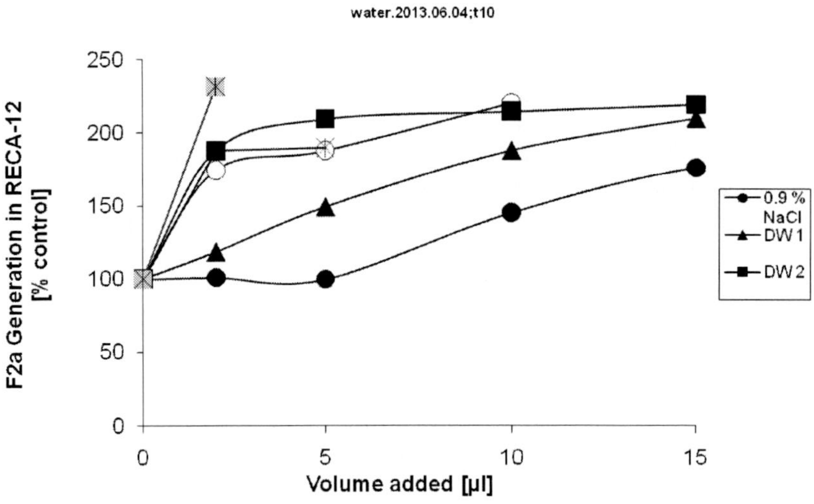

Figure 1. Comparison of different beverages on recalcified thrombin generation. 0-15 µl 0.9% NaCl (•), drinking water 1 (▲), drinking water 2 (■), H_2O (O), Pepsi Cola light (*), or Pepsi Cola (grey*) were added to 40 µl pooled normal citrated plasma. The recalcified coagulation activity assay (RECA) was performed as described under Methods. Intra-assay CV values < 10%.

Glucose triggers altered matrix – coagulation [7-12]. Presumably therefore Pepsi Cola had the worst thrombin generation pattern. The thrombin generation decreased greatly in Pepsi Cola light (without glucose action). Drinking water 1 was much better than drinking water 2; at 2 µl supplementation DW1 had 118% thrombin generation, whereas DW2 had 188% thrombin generation. This thrombin activity was comparable to that induced by Pepsi Cola light (assuming that the beverage is intestinally totally absorbed).

In conclusion, the present easy technique allows to analyze the quality of drinking water or beverages with respect to intrinsic thrombin generation. Solutions similar to 0.9% NaCl are of clinical interest in (bleeding-unrelated) hypovolemic shock [4,5].

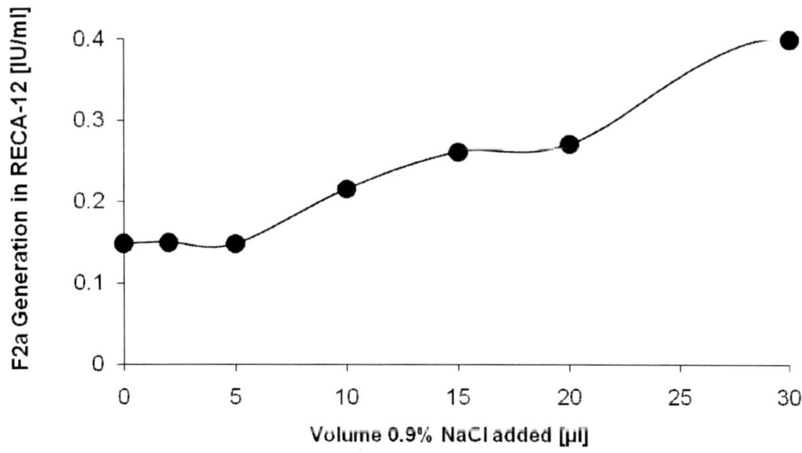

Figure 2. 0.9% NaCl control. 0-30 µl 0.9% NaCl were added to 40 µl pooled normal citrated plasma. The recalcified coagulation activity assay (RECA) was performed as described under Methods.

REFERENCES

[1]   Stief T, Albert A. Water quality routinely tested by RECA. *Hemostasis Laboratory* 2013, 6: 361-5.

[2]   Stief TW. Pathological thrombin generation by the synthetic inhibitor argatroban. *Hemostasis Laboratory* 2009; 2: 83-104.

[3]   Mosesson MW. Update on antithrombin I (fibrin). *Thromb Haemost.* 2007; 98: 105-8.

[4]   Stief TW. Hydroxy-Ethyl-Starch (HES) triggers plasmatic thrombin generation. *Hemostasis Laboratory* 2008; 1: 179-84.

[5]   Stief TW. Therapeutic human albumins trigger plasmatic thrombin generation. *Hemostasis Laboratory* 2008; 1: 185-92.

[6]   Stief TW. Factor 12 activation in two purified systems. *Hemostasis Laboratory* 2011; 4: 385-408.

[7]   Stief TW. Zn^{2+}, hexane, or glucose activate factor 12 and/or prekallikrein in two purified systems. *Hemostasis Laboratory* 2011; 4: 409-26.

[8]   Stief TW. Glucose triggers thrombin generation. *Hemostasis Laboratory* 2010; 3: 93-103.

[9] Stief TW. Glucose activates the early phase of intrinsic coagulation. *Hemostasis Laboratory* 2012; 5: 67-81.

[10] Stief TW, Mohrez M. Glucose activates human intrinsic coagulation *in vivo*. Hemostasis Laboratory 2012; 5: 83-9.

[11] Stief TW, Mohrez M. Glucose provokes pathologic plasmatic thrombin generation. In: *Thrombin: function and pathophysiology*. Stief T, ed.; NOVA science publishers; New York; 2012; pp. 23-35.

[12] Stief TW. The maximal plasma concentration of (delta-)negatively charged contact triggers influences plasmatic thrombin generation. In: *Thrombin: function and pathophysiology*. Stief T, ed.; NOVA science publishers; New York; 2012; pp. 37-46.

Chapter 12

THE TRUE PAP CONCENTRATION IN PLASMA

ABSTRACT

Background: Plasmin-antiplasmin (PAP) is a biomarker for systemic fibrinolysis activation. The reliable quantification of PAP by enzyme immuno assay (EIA) diagnoses the state of systemic fibrinolysis activation. PAP tests are troublesome because fibrinolysis is extremely difficult to stabilize completely. Whereas coagulation is routinely stabilized by withdrawal of calcium ions and arginine concentrations > 200 mM, fibrinolysis is activated by Ca^{2+} withdrawal and arginine is needed at final concentrations > 600 mM. Here PAP is determined for the first time under completely stabilizing conditions.

Material and Methods: 2.6 ml freshest normal venous blood containing 1.6 mg/ml K_3-EDTA after blood withdrawal was immediately mixed with

a) 1.3 ml 1.5 M arginine, pH 8.7 (final plasmatic arginine conc. = 728 mM),
b) 0.337 ml 1.5 M arginine, pH 8.7 (final plasmatic arginine conc. = 315 mM),
c) control 1 (final plasmatic arginine conc. = 0 mM),
d) control 2 (final plasmatic arg. conc. = 1250 mM added 1h after blood withdrawal).

10 µl sample were added in triplicate to the (first monoclonal PAP-6) antibody against the antiplasmin neoantigen coated microwells of the PAP- enzyme immune assay (EIA) from DRG, Marburg, Germany that had been prefilled with 50 µl assay buffer and 50 µl arginine, pH 8.7 (final conc. 1.25 M). After 1h (23°C) the plate was washed 3fold, the (second polyclonal) antibody directed against plasmin(ogen) coupled with horse radish peroxidase was added. After 1h (23°C) the plate was washed 3fold and tetramethylbenzidine was added. The reaction was followed at 405 nm, after 1h (23°C) a sufficiently high absorbance had been generated to stop the reaction with 0.5 N H_2SO_4 and to read the absorbance at 450 nm.

Results and Discussion: The PAP concentrations were (hematocrit corrected) 1.6 ng/ml in 728 mM arginine EDTA plasma, 4.8 ng/ml in 315 mM arginine EDTA plasma, 11.2 ng/ml in 0 mM arginine EDTA plasma. The second control plasma with 1.25 M arginine stabilization 1h after blood withdrawal had 4.9 ng/ml PAP. To obtain the true *in vivo* basal PAP conc. EDTA blood must immediately be stabilized by at least about 700 mM arginine and the EIA should be performed with 1.25 M arginine in the antigen-capture phase. The true 100% normal basal PAP concentration seems to be very close to the true 100% normal basal TAT concentration that is about 1.3 ng/ml.

INTRODUCTION

Hemostasis activation is more important than hemostasis (the system of generation and destruction of thrombi). A diagnosis of hemostasis activation within the first six golden hours of a disease enables the rapid adequate therapy [1-3]. Coagulation activation is diagnosed by measuring the systemic biomarker F2a···α_2M (thrombin entrapped in α_2-macroglobulin) that allows the NIC···*PIC-0*···*PIC-1*···*PIC-2* classification, the initial respective PIC stages beginning after *120···150···200%* of normal. Fibrinolysis activation can similarly be staged as NIF···*PIF-0*···*PIF-1*···*PIF-2*. Pli···α_2M (plasmin entrapped in α_2-macroglobulin) is probably also an excellent biomarker for hemostasis activation. However due the extremely efficient plasmin inactivation of antiplasmin (plasmin-inhibitor = PI; 0.1-1% of systemic plasmin ends in α_2M) when compared with that of antithrombin-3 (AT-3) against thrombin (about 10% of systemic thrombin ends here) and to slower cleavage of the chromogenic plasmin substrate compared with a fast chromogenic thrombin substrate the current state of photometer art allows systemic thrombin activity determinations within 1h (37°C) but for systemic plasmin activity determinations at least 3d (37°C) are necessary. Therefore, Pli···α_2M for the moment is not a suitable routine biomarker of fibrinolysis activation. Instead, plasmin-antiplasmin (PAP) could be measured as a biomarker for systemic fibrinolysis activation. However, enzyme-inhibitor complexes (PAP or TAT) tend to "jump" in the daily clinical monitoring of a patient [4,5].

MATERIAL AND METHODS

2.6 ml freshest normal venous blood of a healthy donor drawn after written informed consent into polypropylene monovettes (Sarstedt, Nümbrecht, Germany) containing 1.6 mg/ml K_3-EDTA was mixed with

a) 1.3 ml 1.5 M arginine, pH 8.7 (final plasmatic arginine conc. = 728 mM),
b) 0.337 ml 1.5 M arginine, pH 8.7 (Sigma, Deisenhofen, Germany) (final plasmatic arginine conc. = 315 mM),
c) control 1 (final plasmatic arginine conc. = 0 mM),
d) control 2 (final plasmatic arg. conc. = 1250 mM added 1h after blood withdrawal).

10 µl sample or standards containing 0-50 ng/ml PAP were added in triplicate to the (first monoclonal PAP-6) antibody against the antiplasmin neoantigen coated microwells of the PAP-enzyme immuno assay (EIA) from DRG, Marburg, Germany that had been prefilled with 50 µl assay buffer and 50 µl arginine, pH 8.7 (final conc. 1.25 M). After 1h (23°C) the plate was washed 3fold, the (second polyclonal) antibody directed against plasmin(ogen) coupled with horse radish peroxidase was added. After 1h (23°C) the plate was washed 3fold and tetramethylbenzidine (TMB) was added. The reaction was followed at 405 nm, after 1h (23°C) a sufficiently high absorbance had been generated to stop the reaction with 0.5 N H_2SO_4 and to read the extinction at 450 nm.

RESULTS AND DISCUSSION

The calibration curves for the inverse functions (PAP conc. → 450 nm absorbance) for low PAP conc. and high PAP conc. are demonstrated in figure 1 and 2, respectively.

The PAP concentrations were (hematocrit corrected) 1.6 ng/ml in 728 mM arginine K_3-EDTA plasma, 4.8 ng/ml in 315 mM arginine K_3-EDTA plasma, 11.2 ng/ml in 0 mM arginine K_3-EDTA plasma (Figure 3). Higher arginine concentrations than 728 mM might result in hemolysis [6,7].

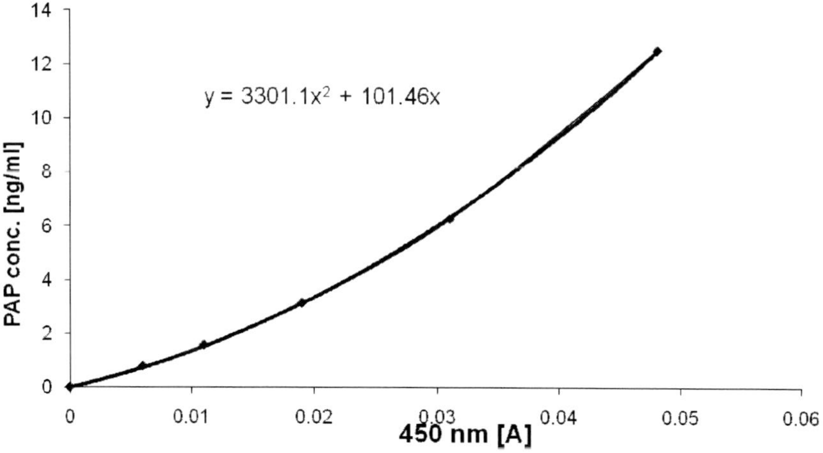

Figure 1. Calibration curve for inverse function of PAP conc. (y) →450 nm absorbance (x). The transformation formula $y = 3301,1x^2 + 101,46x$ is valid for PAP concentrations up to 12.5 ng/ml.

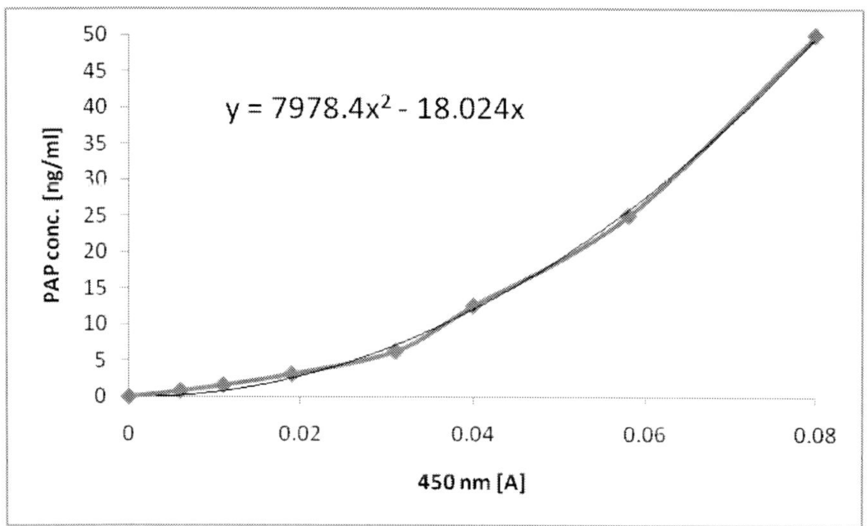

Figure 2. Calibration curve for inverse function of PAP conc. (y) →450 nm absorbance (x). The transformation formula $y = 7978,4x^2 - 18,024x$ is valid for PAP concentrations of 12.5-50 ng/ml.

The second control plasma with stabilization by 1.25 M arginine 1h after blood withdrawal had 4.9 ng/ml PAP. This means that to obtain the true *in vivo* basal PAP conc. EDTA blood must immediately be stabilized by at least about 700 mM arginine [7,8]. The true 100% normal basal PAP concentration seems to be very close to the 100% normal basal TAT concentration that is about 1.3 ng/ml [9,10]. In plasma plasmin can also be inhibited by fragment D- inhibitor, an inhibitor comparable to antithrombin-1 for thrombin [11-21].

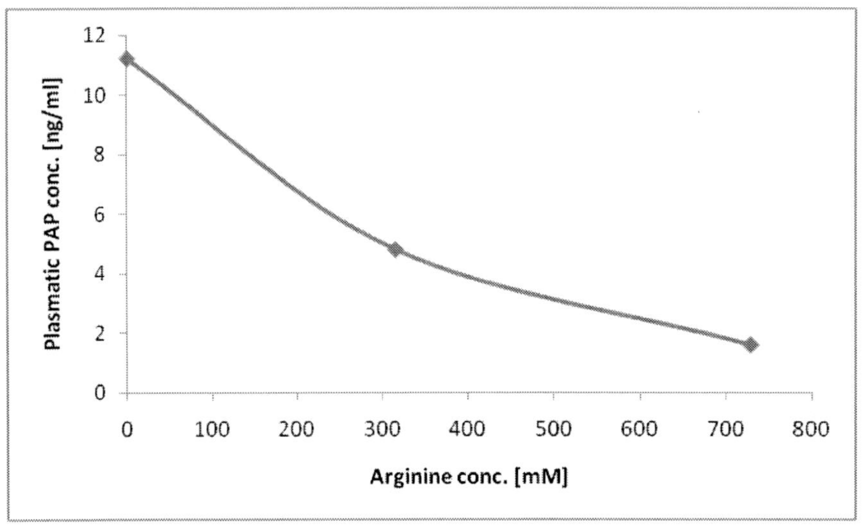

Figure 3. Stabilization of basal plasmatic fibrinolysis by arginine. Without arginine addition 11.2 ng/ml PAP were measured in EDTA plasma. In presence of 728 mM arginine the plasmatic PAP conc. was only 1.6 ng/ml.

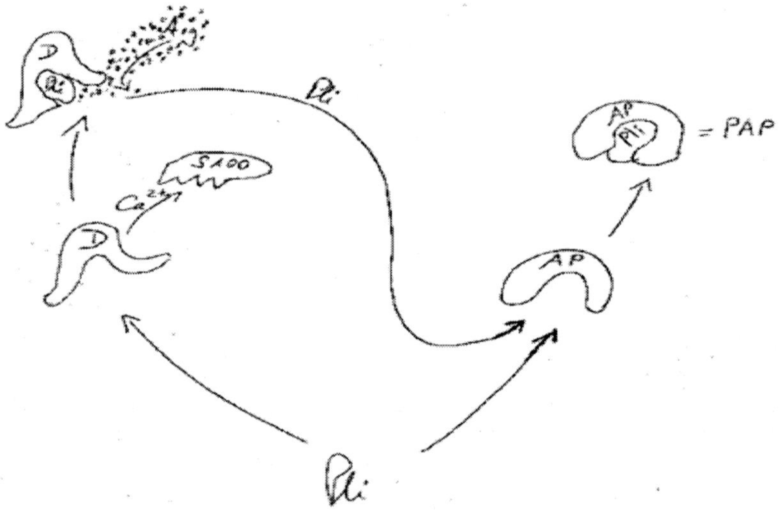

Figure 4. Schematic interaction of plasmin with antiplasmin or fragment D [11,12]. Plasmin (Pli) is inactivated by antiplasmin (AP) or inhibited by Ca^{2+} deficient fibrinogen fragment D. Chaotropic compounds as arginine can solve the Pli-D binding (similar to that between antithrombin-1/thrombin) and direct Pli towards its specific inactivator antiplasmin [11-21].

REFERENCES

[1] van Gorp EC, Minnema MC, Suharti C, Mairuhu AT, Brandjes DP, ten Cate H, Hack CE, Meijers JC. Activation of coagulation factor XI, without detectable contact activation in dengue haemorrhagic fever. *Br J Haematol.* 2001; 113: 94-9.

[2] Stief TW. The laboratory diagnosis of the pre-phase of pathologic disseminated intravascular coagulation. *Hemostasis Laboratory* 2008; 1: 2-20.

[3] Stief TW. Thrombin – applied clinical biochemistry of the main factor of coagulation. In: *Thrombin: function and pathophysiology*. Stief T, ed.; Nova science publishers; New York; 2012; pp. vii-xx.

[4] Stief TW, Ulbricht K, Max M. Circulating thrombin activity in sepsis. *Hemostasis Laboratory* 2009; 2: 293-306.

[5] Stief TW, Ulbricht K, Max M. Circulating plasmin activity in severe sepsis. *Hemostasis Laboratory* 2010; 3: 105-20.

[6] Stief TW, Richter A, Bünder R, Maisch B, Renz H. Functional determination of plasmin
in arginine-stabilized plasma. *Clin Appl Thrombosis/Hemostasis* 2005; 11: 303-310.

[7] Stief TW, Richter A, Bünder R, Maisch B, Renz H. Monitoring of plasmin- and plasminogen-activator-activity in blood of patients under fibrinolytic treatment by reteplase. *Clin Appl Thromb/Hemostasis* 2006; 12: 213-18.

[8] Stief TW. Arginine conserves the hemostasis activation state of plasma even against freezing/thawing. In: *Diamino Amino Acids*; Schäfer HA, Wohlbier LM, eds; Nova Biomedical, New York, 2008; pp. 219-33.

[9] van der Bom JG, Bots ML, Haverkate F, Meijer P, Hofman A, Kluft C, Grobbee DE. Activation products of the haemostatic system in coronary, cerebrovascular and peripheral arterial disease. *Thromb Haemost.* 2001; 85: 234-9.

[10] Stief TW, Ijagha O, Weiste B, Herzum I, Renz H, Max M. Analysis of hemostasis alterations in sepsis. *Blood Coagul Fibrinolysis* 2007; 18: 179-86.

[11] Suenson E, Thorsen S. Secondary-site binding of Glu-plasmin, Lys-plasmin and miniplasmin to fibrin. *Biochem J.* 1981; 197: 619-28.

[12] Suenson E, Bjerrum P, Holm A, Lind B, Meldal M, Selmer J, Petersen LC. The role of fragment X polymers in the fibrin enhancement of tissue plasminogen activator-catalyzed plasmin formation. *J Biol Chem.* 1990; 265: 22228-37.

[13] Stief TW, Lenz P, Becker U. A simple method for producing degradation products of fibrinogen by an insoluble derivative of plasmin. *Thromb Res.* 1987; 48: 603-9.

[14] Bosma PJ, Rijken DC, Nieuwenhuizen W. Binding of tissue-type plasminogen activator to fibrinogen fragments. *Eur J Biochem.* 1988; 172: 399-404.

[15] Stief TW, Marx R, Heimburger N. Oxidized fibrin(ogen) derivatives enhance the activity of tissue type plasminogen activator. *Thromb Res.* 1989; 56: 221-8.

[16] Stief TW, Kretschmer V, Kosche B, Doss MO, Renz H. Thrombin converts singlet oxygen ($^1O_2$)-oxidized fibrinogen into a soluble t-PA cofactor. A new method for preparing a stimulator for functional t-PA assays. *Ann Hematol.* 2001; 80: 189-94.

[17] Haverkate F, Timan G. Protective effect of calcium in the plasmin degradation of fibrinogen and fibrin fragments D. *Thromb Res.* 1977; 10: 803-12.

[18] Perera C, McNeil HP, Geczy CL. S100 Calgranulins in inflammatory arthritis. *Immunol Cell Biol.* 2010; 88:v41-9.

[19] Stief TW. Inhibition of thrombin in plasma by heparin or arginine. *Clin Appl Thrombosis/Hemostasis* 2007; 13: 146-153.

[20] Stief TW. Thrombin generation by hemolysis. *Blood Coagulation & Fibrinolysis* 2007; 18: 61-66.

[21] Mosesson MW. Update on antithrombin I (fibrin). *Thromb Haemost.* 2007; 98: 105-8.

Chapter 13

STIMULATION OF BLOOD SINGLET OXYGEN GENERATION BY SERINE PROTEASES

ABSTRACT

Background: Proteases are very important regulators of cell function. The up-regulation of the NADPH-oxidase, the physiologic generator system of large quantities of reactive oxygen species (ROS; H_2O_2 being the primary ROS and $^1\Delta O_2^*$ being the most selective ROS against "non-self") is the central activity of neutrophils upon contact with fungi, bacteria, or micro-thrombi. The regulation of $^1\Delta O_2^*$ generation by serine proteases is the purpose of this work.

Material and Methods: Freshest citrated blood or EDTA-blood (10 µl) were incubated with 120 µl Hanks´ Balanced Salt Solution (HBSS without phenol red), 30 µl urokinase, t-PA, plasmin, factor 12a, kallikrein, trypsin, factor 9a, factor 11a, factor 10a, or thrombin in 5% human albumin, 10 µl 0.28 mM luminol (final conc.), 10 µl 2 µg/ml zymosan A (final conc.) in black polystyrene plates (Brand®781608). The luminescence was measured in the ascending part of the ROS generation kinetic by a photons-multiplying microtiter plate luminometer with an integration time of 0.5s per well. Citrated blood was also analyzed in the blood ROS generation assay with 60 min pre-incubation (BRGA-60-).

Results and Discussion: Urokinase stimulated blood ROS generation about 5fold stronger than t-PA in citrated blood. The approx. SC200 were 17 IU/ml u-PA, approx. SC150 = 10 IU/ml u-PA or 300 IU/ml t-PA (6.3 IU t-PA are equivalent to 1 IU u-PA). The approx. SC200 for plasmin in citrated blood was 1.2 mIU/ml. The approx. SC200 was 2.5 mIU/ml F9a, 10 ng/ml trypsin, or 0.2 ng/ml thrombin. Kallikrein inhibited the singlet oxygen generation with an approx. IC50 of 3 ng/ml and F12a had an approx. IC50 of 19 ng/ml.

In EDTA-blood the 200% stimulatory concentrations were 30 IU/ml u-PA, 300 IU/ml t-PA, 2.5 mIU/ml plasmin, 0.3 mIU/ml F9a, 12 mUml F10a, 10 ng/ml F11a, 10 ng/ml trypsin, 0.6 mIU/ml thrombin, 35 ng/m F12a, 4 ng/ml kallikrein.

The approx. SC200 values were in the BRGA-60- version of the assay (citrated blood) were 4 IU/ml urokinase, 2.5 mIU/ml plasmin, 15 ng/ml trypsin, 3 ng/ml kallikrein, 6 ng/ml F12a, 6 mIU/ml F9a, 3 mU/ml F10a, or 0.5 mIU/ml F2a.

All serine proteases seem to upregulate the NADPH-oxidase assembly. Only kallikrein and F12a, the starter enzymes of intrinsic coagulation, in the initial phase of neutrophil activation behaved as inhibitors of NADPH-oxidase assembly.

Keywords: Serine proteases, reactive oxygen species, ROS, singlet oxygen, NADPH-oxidase, neutrophils

INTRODUCTION

Serine proteases with their catalytic His···Asp···Ser center are very important regulators of cell function. The up-regulation of the NADPH-oxidase, the physiologic generator system of large quantities of reactive oxygen species (ROS; H_2O_2 being the primary ROS and $^1\Delta O_2^*$ being the most selective ROS against "non-self") is the central activity of neutrophils upon contact with fungi, bacteria, or micro-thrombi. The regulation of $^1\Delta O_2^*$ generation by serine proteases [1-17] is the purpose of this work.

MATERIAL AND METHODS

10 µl freshest (less than 0.5h old) individual normal venous blood of a healthy donor drawn from a left hand vein into polypropylene monovettes (4.5 ml blood added to 0.5 ml 106 mM sodium citrate, pH 7.4 or 2.6 ml blood supplemented with 1.6 mg/ml K_3-EDTA; Sarstedt, Nümbrecht, Germany) were incubated in duplicate) in black polystyrene plates (Brand, Wertheim, Germany; article nr. 781608) with 120 µl Hanks´ Balanced Salt Solution (HBSS without phenol red; Sigma, Deisenhofen, Germany), 30 µl 0-92 ng/ml (final conc.) human plasma kallikrein (Sigma; article nr. K2638-50UG; stem solution 50 µg/ml in 29% glycerol) in 5% human albumin (CSL Behring, Marburg, Germany), 0-833 IU/ml (final conc.) tissue type plasminogen activator (t-PA; Boehringer Ingelheim, Germany), 0-132 IU/ml (final conc.) urokinase (medac, Hamburg, Germany), 0-17 mIU/ml plasmin (3rd international plasmin standard; NIBSC, Potters Bar, England) 0-18 mIU/ml factor 9a (NIBSC,), 0-70 mU/ml bovine factor 10a (Chromogenix, Mölndal, Sweden), 0-1100 ng/ml factor 11a (Enzyme Research Laboratories– Haemochrom, Essen, Germany), 0-112 ng/ml porcine trypsin (Sigma), or 0-148 ng/ml factor 12a (Enzyme Research Laboratories– Haemochrom).

After 0 min (BRGA) or 60 min (BRGA-60-) 10 µl 0.28 mM luminol (final conc.) (Sigma) and 10 µl 2 µg/ml zymosan A (final conc.) (Sigma) were added. The luminescence in the ascending part of the ROS generation kinetic was measured by a photons-multiplying microtiter plate luminometer (LUmo; Autobio-anthos, Krefeld, Germany) with an integration time of 0.5s per well [17,18]. The approximate 200% stimulatory or 50% inhibitory concentrations of the serine proteases on blood ROS generation were determined (approx. SC200, approx. IC50).

RESULTS AND DISCUSSION

Figures 1-19 demonstrate the modulation of the generation of singlet oxygen by serine proteases in the BRGA. Figures 20-23 represent the data of the BRGA-60-.

Urokinase stimulated blood ROS generation about 5fold stronger than t-PA in citrated blood. The approx. SC200 were 17 IU/ml u-PA, approx. SC150 = 10 IU/ml u-PA or 300 IU/ml t-PA (6.3 IU t-PA are equivalent to 1 IU u-PA) (Figures 1,2). The approx. SC200 for plasmin in citrated blood was 1.2 mIU/ml (Figure 3). The approx. SC200 was 2.5 mIU/ml F9a (Figure 4), 10 ng/ml trypsin (Figure 6), or 0.2 ng/ml thrombin (Figure 9). Kallikrein inhibited the singlet oxygen generation with an approx. IC50 of 3 ng/ml (Figure 7) and F12a had an approx. IC50 of 19 ng/ml (Figure 8).

In EDTA-blood the 200% stimulatory concentrations were 30 IU/ml u-PA, 300 IU/ml t-PA, 2.5 mIU/ml plasmin, 0.3 mIU/ml F9a, 12 mUml F10a, 10 ng/ml F11a, 10 ng/ml trypsin, 0.6 mIU/ml thrombin. 35 ng/m F12a, 4 ng/ml kallikrein. The EDTA-blood results can vary from the citrated blood results because EDTA chelates Ca^{2+} stronger than citrate (imitating the S100 family) and fibrin subunit D generated in Ca^{2+} absence is different to the product generated in its presence [19].

The approx. SC200 values were in the BRGA-60- version of the assay (citrated blood) were 4 IU/ml urokinase, 2.5 mIU/ml plasmin, 15 ng/ml trypsin, 3 ng/ml kallikrein, 6 ng/ml F12a, 6 mIU/ml F9a, 3 mU/ml F10a, or 0.5 mIU/ml F2a.

Thus, any serine protease seems to upregulate the NADPH-oxidase assembly [20]. Only kallikrein and F12a, the starter enzymes of intrinsic coagulation, in the initial phase of neutrophil activation behaved as inhibitors of NADPH-oxidase assembly.

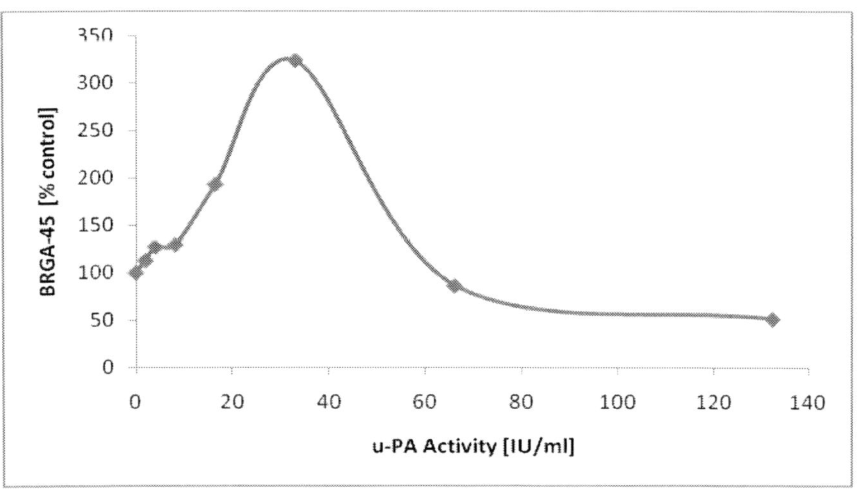

Figure 1. Stimulation of blood ROS generation by urokinase. The blood ROS generation assay (BRGA) was performed with freshest citrated blood in presence of 0-132 IU/ml urokinase as described in Methods. The approx. SC200 was 17 IU/ml urokinase (approx. 150= 10 IU/ml u-PA). The ROS maximum (about 2000 RLU/s) appeared at about 80 min (37°C; the 100% control value of BRGA-45 was about 400 RLU/s).

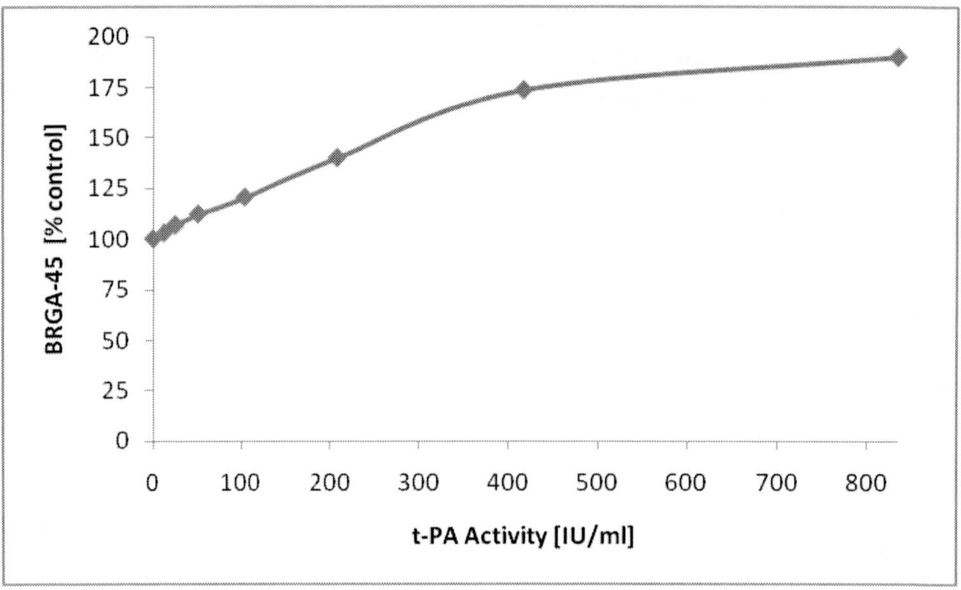

Figure 2. Stimulation of blood ROS generation by t-PA. The blood ROS generation assay (BRGA) was performed in presence of 0-833 IU/ml t-PA as described in Methods. The approx. SC150 was 300 IU/ml t-PA.

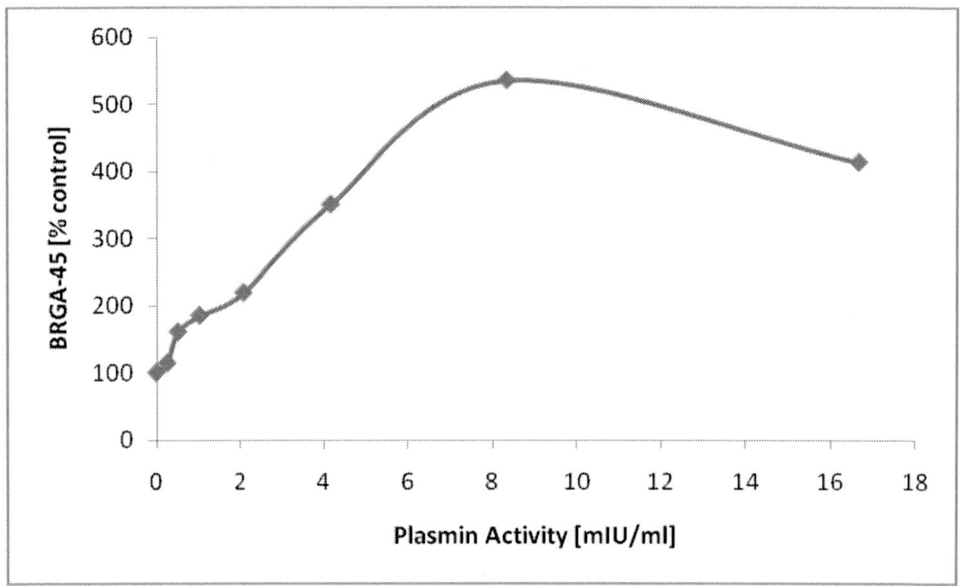

Figure 3. Stimulation of blood ROS generation by plasmin. The blood ROS generation assay (BRGA) was performed with freshest citrated blood in presence of 0-17 mIU/ml human plasmin as described in Methods. The approx. SC200 was 1.2 mIU/ml plasmin.

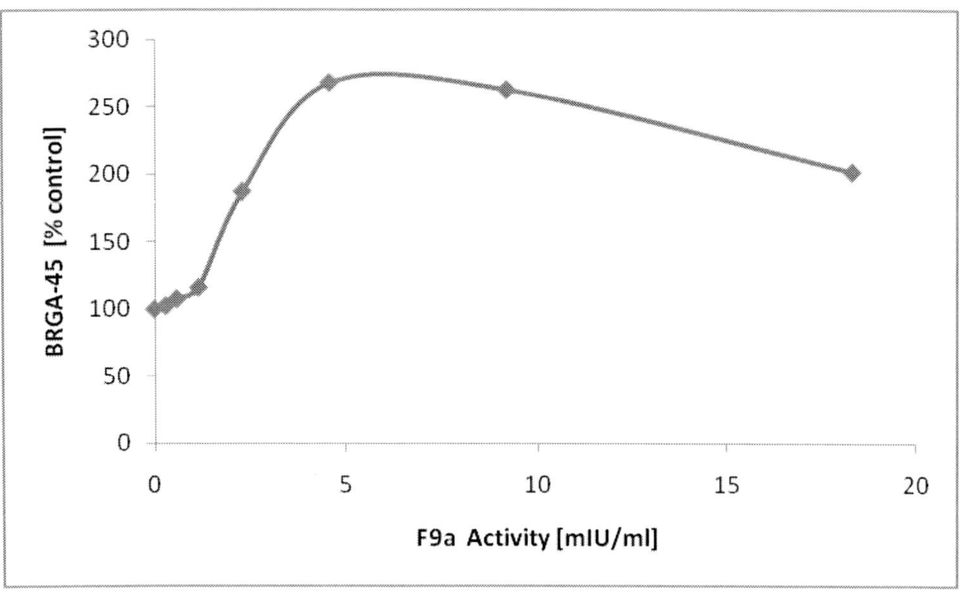

Figure 4. Stimulation of blood ROS generation by F9a. The BRGA was performed with freshest citrated blood in presence of 0-18 mIU/ml human factor 9a as described in Methods. The approx. SC200 was 2.5 mIU/ml F9a.

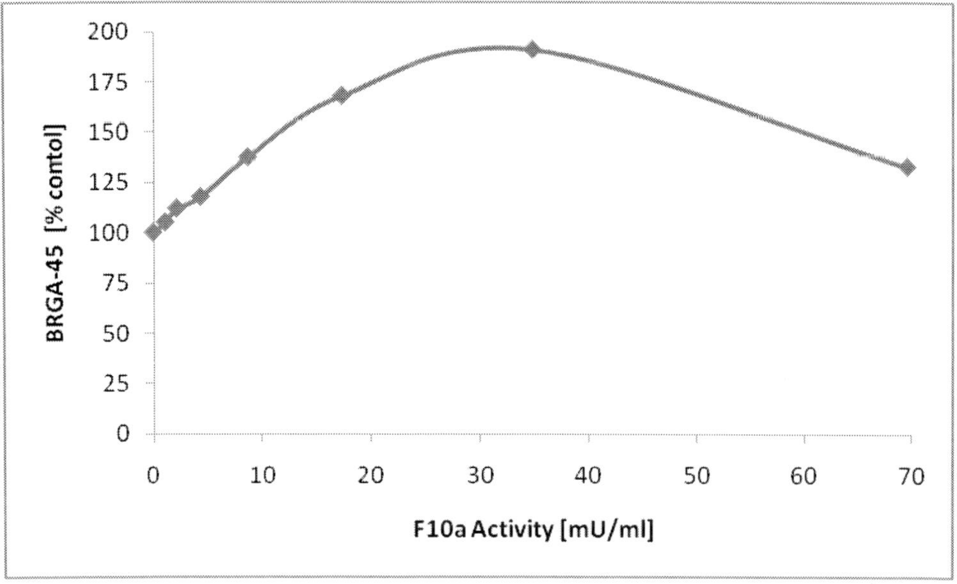

Figure 5a. Stimulation of blood ROS generation by factor 10a. The BRGA was performed with freshest citrated blood in presence of 0-70 mU/ml factor 10a as described in Methods. The approx. SC150 was 11 mU/ml F10a.

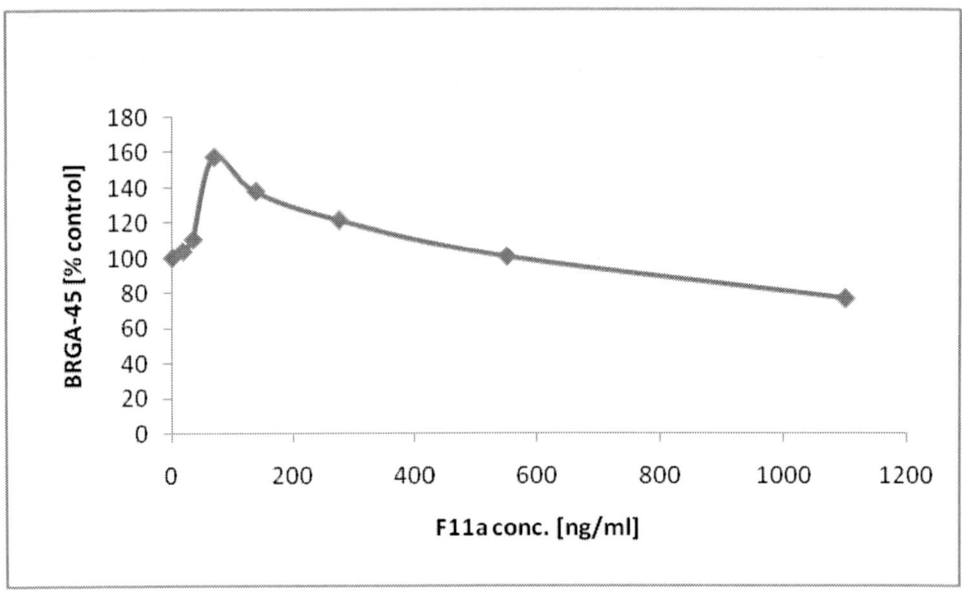

Figure 5b. Stimulation of blood ROS generation by factor 11a. The BRGA was performed with freshest citrated blood in presence of 0-1100 ng/ml factor 11a as described in Methods. The approx. SC150 was 60 ng/ml F11a.

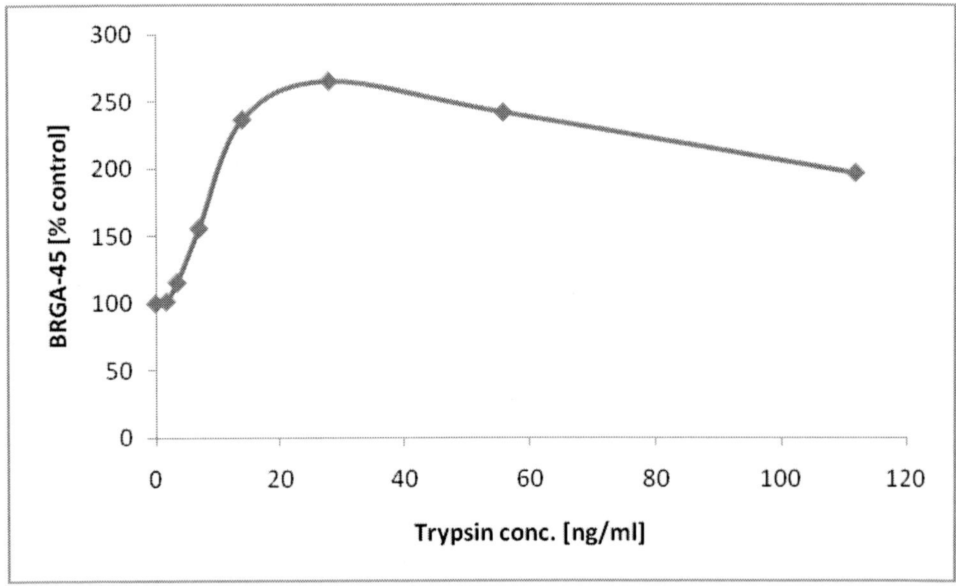

Figure 6. Stimulation of blood ROS generation by trypsin. The BRGA was performed with freshest citrated blood in presence of 0-112 ng/ml trypsin as described in Methods. The approx. SC200 was 10 ng/ml trypsin.

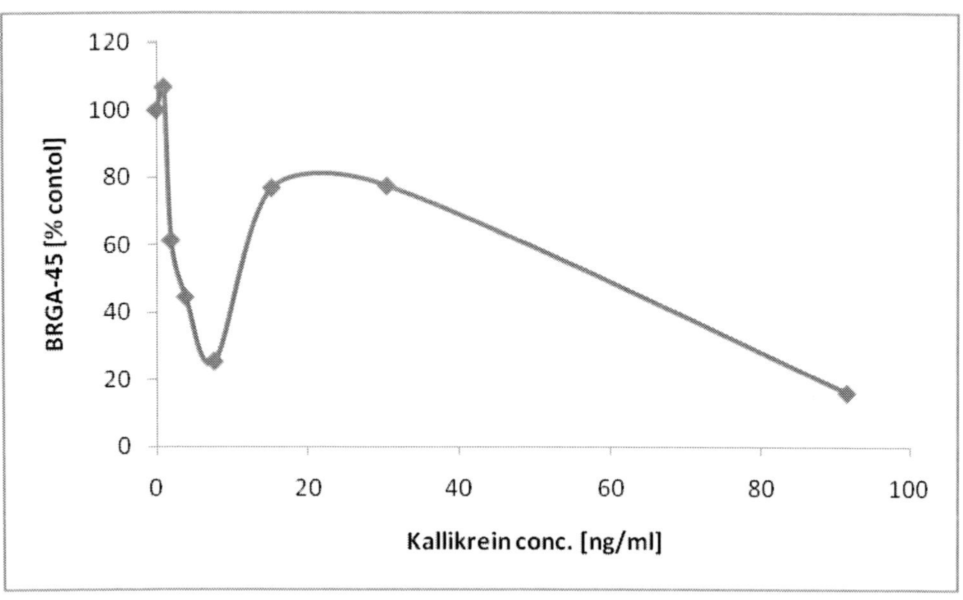

Figure 7. Inhibition of blood ROS generation by kallikrein. The BRGA was performed with freshest citrated blood in presence of 0-92 ng/ml human plasma kallikrein as described in Methods. The approx. IC50 was 3 ng/ml kallikrein.

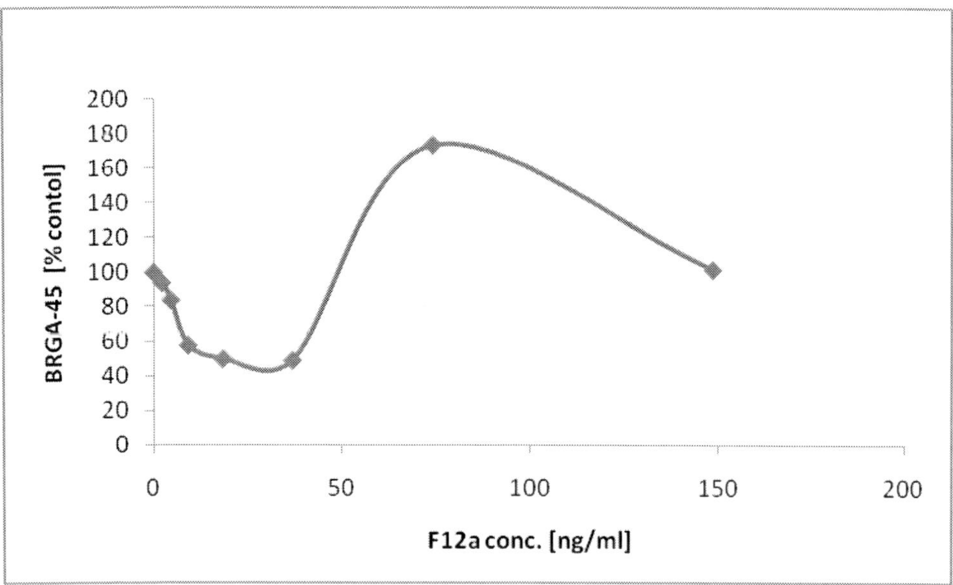

Figure 8. Inhibition of blood ROS generation by factor 12a. The BRGA was performed with freshest citrated blood in presence of 0-148 ng/ml factor 12a as described in Methods. The approx. IC50 was 19 ng/ml F12a.

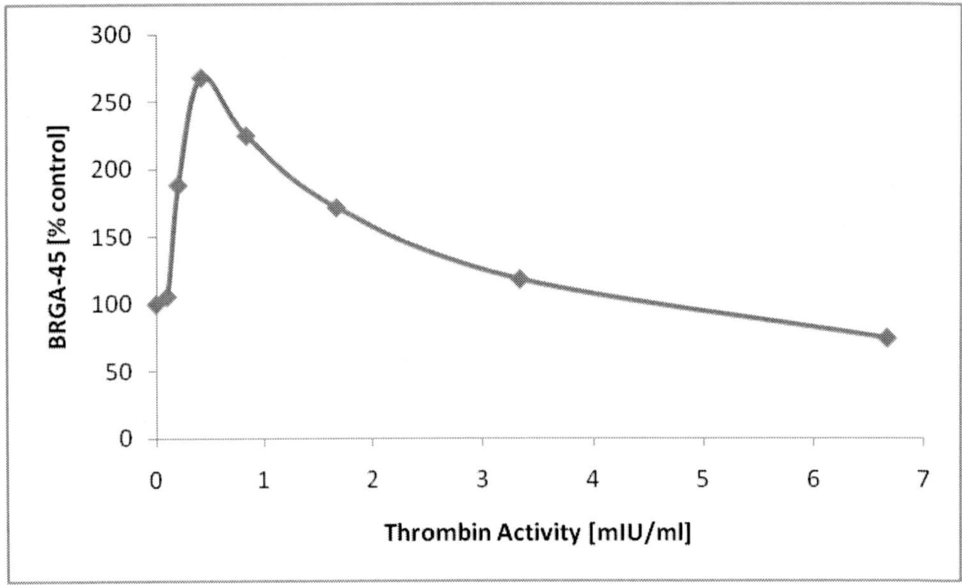

Figure 9. Stimulation of blood ROS generation by thrombin. The BRGA was performed with freshest citrated blood in presence of 0-6.7 mIU/ml bovine thrombin as described in Methods. The approx. SC200 was 0.2 mIU/ml thrombin.

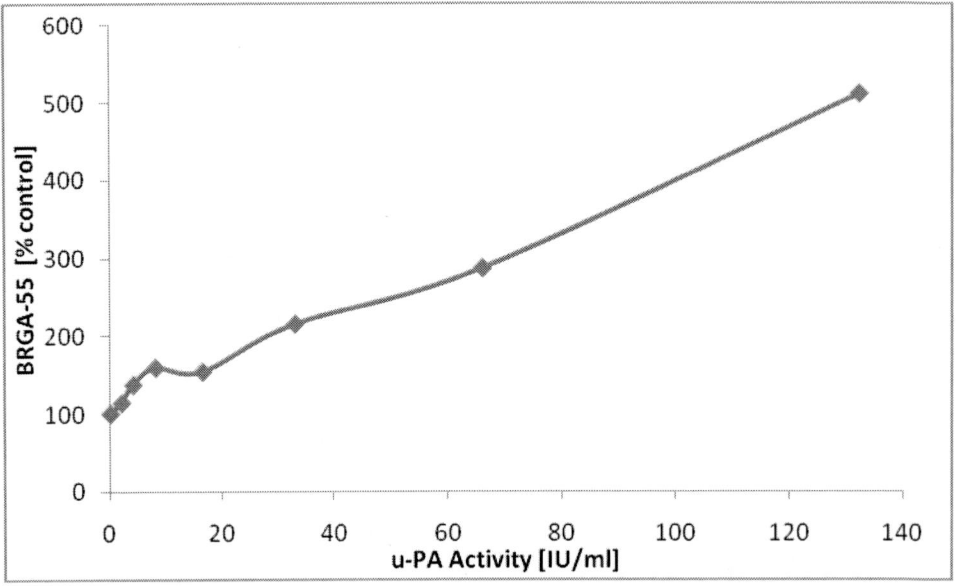

Figure 10. Stimulation of blood ROS generation by urokinase. The BRGA was performed with freshest EDTA-blood in presence of 0-132 IU/ml urokinase as described in Methods (100% control = about 800 RLU/s). The approx. SC200 was 30 IU/ml u-PA.

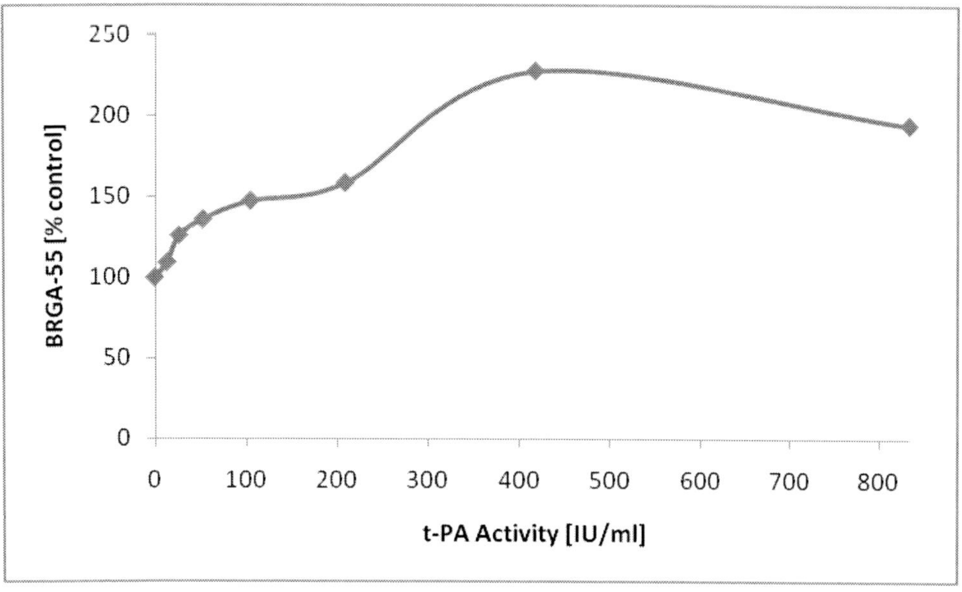

Figure 11. Stimulation of blood ROS generation by t-PA. The BRGA was performed with freshest EDTA-blood in presence of 0-833 IU/ml t-PA as described in Methods. The approx. SC200 was 300 IU/ml t-PA.

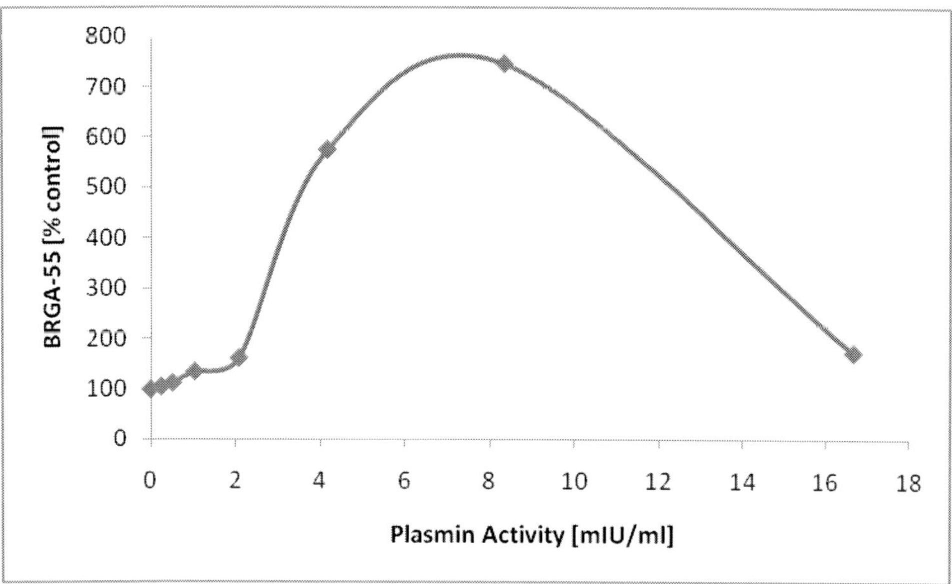

Figure 12. Stimulation of blood ROS generation by plasmin. The BRGA was performed with freshest EDTA-blood in presence of 0-17 mIU/ml plasmin as described in Methods. The approx. SC200 was 2.5 mIU/ml plasmin.

Thomas Stief

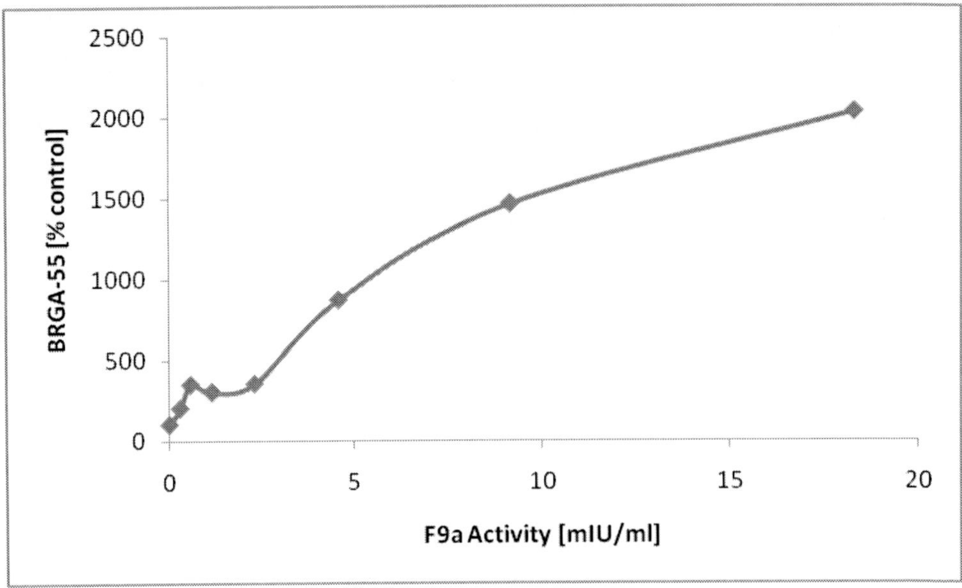

Figure 13. Stimulation of blood ROS generation by factor 9a. The blood ROS generation assay (BRGA) was performed with freshest EDTA-blood in presence of 0-18 mIU/ml factor 9a as described in Methods. The approx. SC200 was 0.3 mIU/ml F9a.

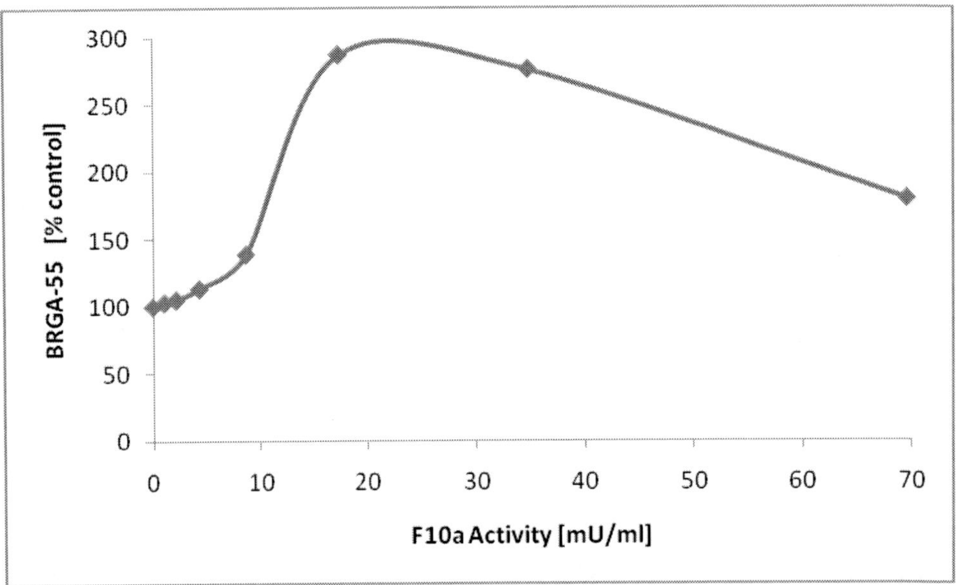

Figure 14. Stimulation of blood ROS generation by factor 10a. The blood ROS generation assay (BRGA) was performed with freshest EDTA-blood in presence of 0-70 mU/ml factor 10a as described in Methods. The approx. SC200 was 12 mU/ml F10a.

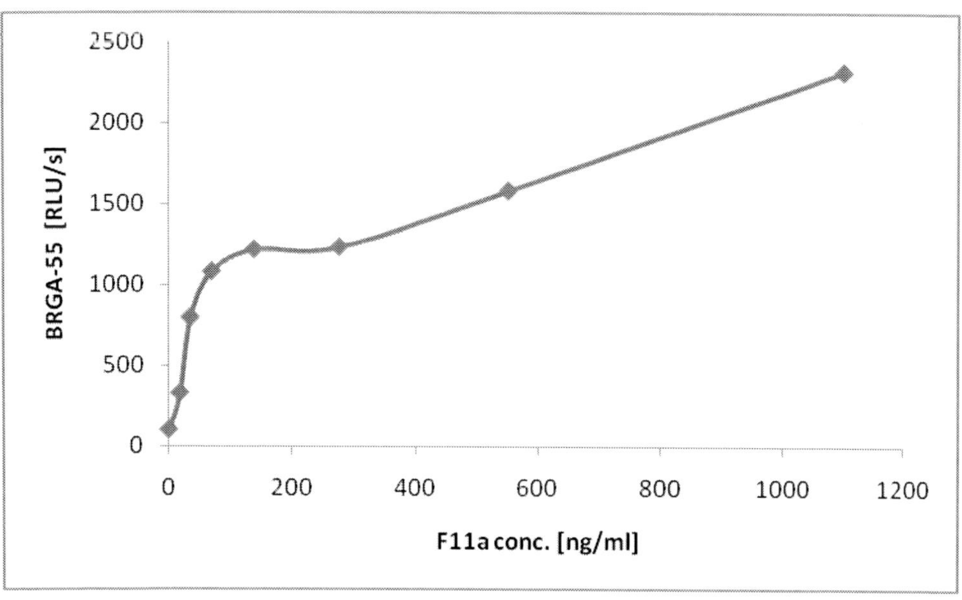

Figure 15. Stimulation of blood ROS generation by factor 11a. The blood ROS generation assay (BRGA) was performed with freshest EDTA-blood in presence of 0-1100 ng/ml factor 11a as described in Methods. The approx. SC200 was 10 ng/ml F11a.

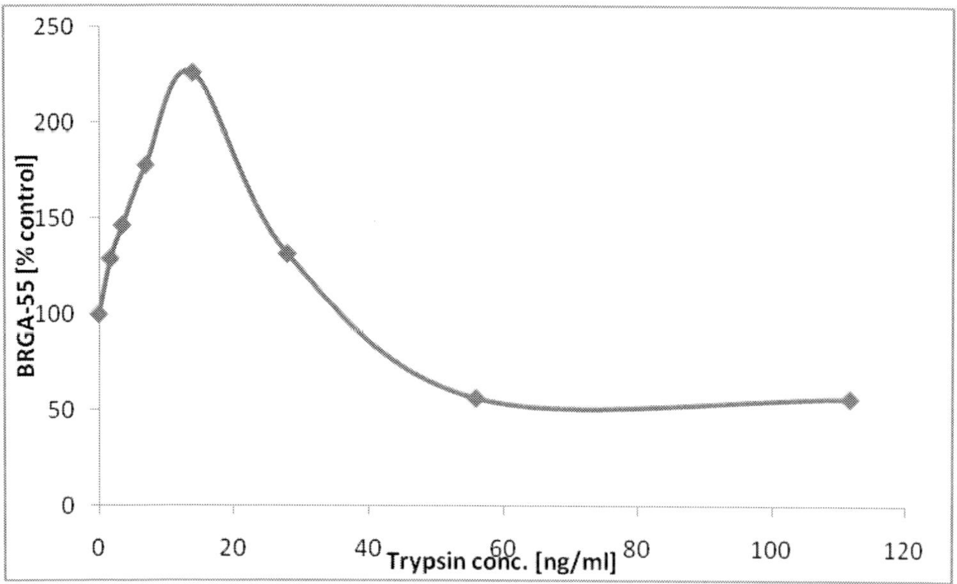

Figure 16. Stimulation of blood ROS generation by trypsin. The blood ROS generation assay (BRGA) was performed with freshest EDTA-blood in presence of 0-112 ng/ml porcine trypsin as described in Methods. The approx. SC200 was 10 ng/ml trypsin.

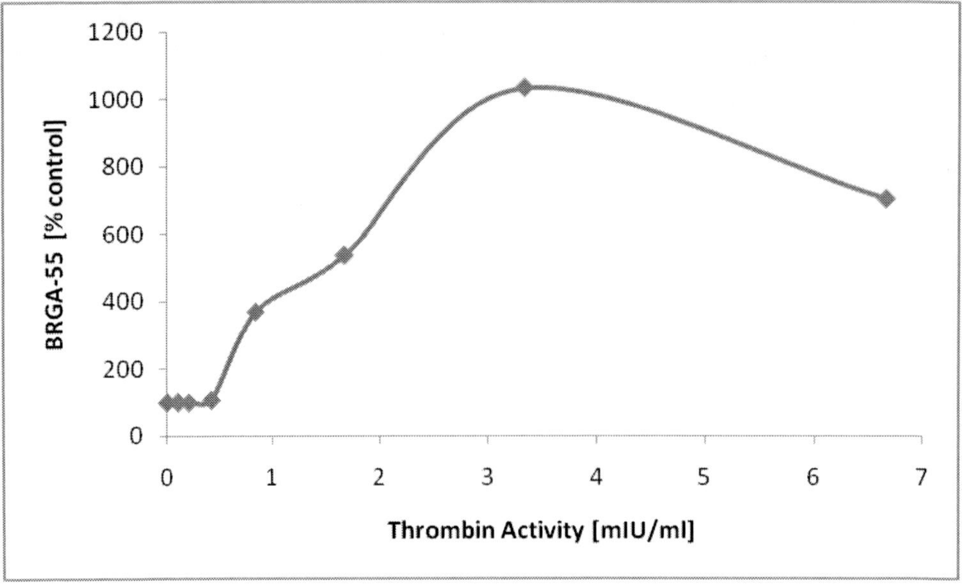

Figure 17. Stimulation of blood ROS generation by thrombin. The blood ROS generation assay (BRGA) was performed with freshest EDTA-blood in presence of 0-6.7 mIU/ bovine thrombin as described in Methods. The approx. SC200 was 0.6 mIU/ml thrombin.

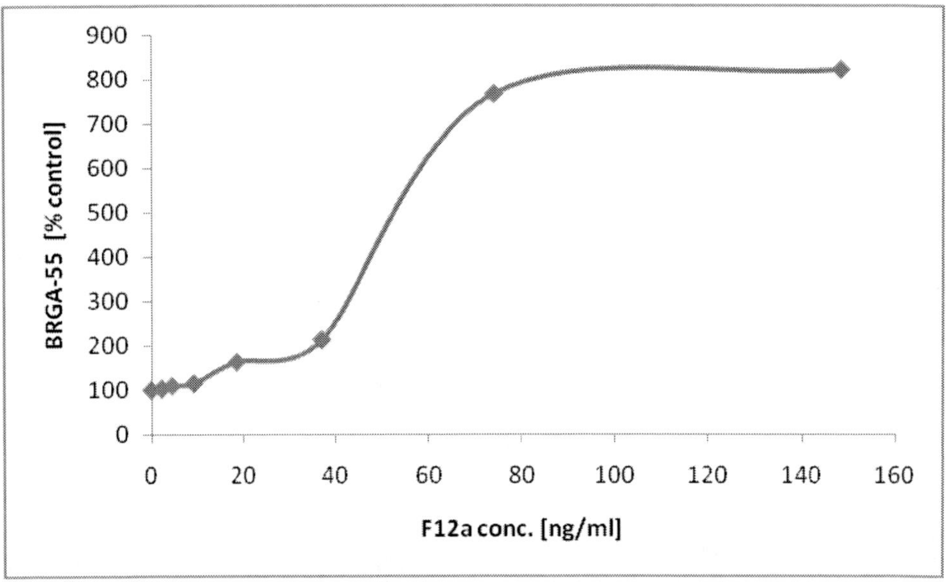

Figure 18. Stimulation of blood ROS generation by F12a. The blood ROS generation assay (BRGA) was performed with freshest EDTA-blood in presence of 0-148 ng/ml F12a as described in Methods. The approx. SC200 was 35 ng/ml F12a.

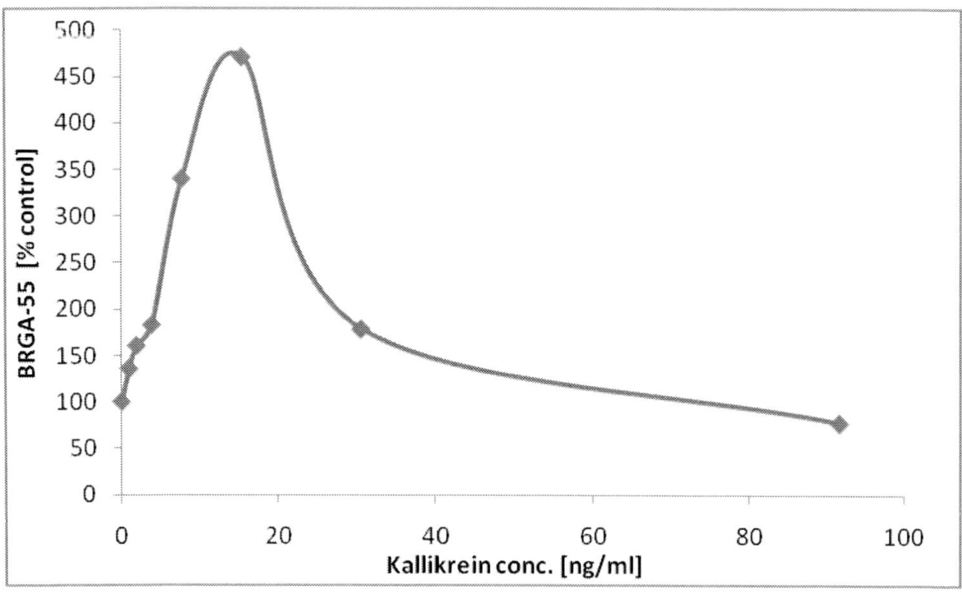

Figure 19. Stimulation of blood ROS generation by kallikrein. The blood ROS generation assay (BRGA) was performed with freshest EDTA-blood in presence of 0-92 ng/ml human plasma kallikrein as described in Methods. The approx. SC200 was 4 ng/ml kallikrein.

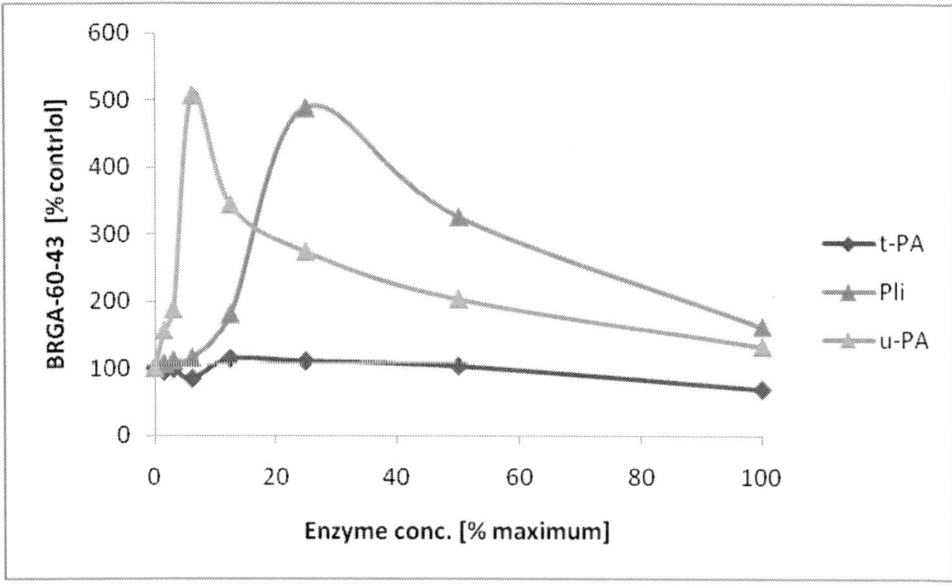

Figure 20. Stimulation of blood ROS generation by u-PA or plasmin. The blood ROS generation assay with 60 min pre-incubation and 43 min incubation time (BRGA-60-43) was performed with freshest citrated blood in presence of 0-833 IU/ml t-PA, 0-17 mIU/ml plasmin, or 0-132 IU/ml u-PA (final conc.) as described in Methods (100% = about 1500 RLU/s). The approx. SC200 values were – IU/ml t-PA, 2.5 mIU/ml plasmin, 4 IU/ml urokinase.

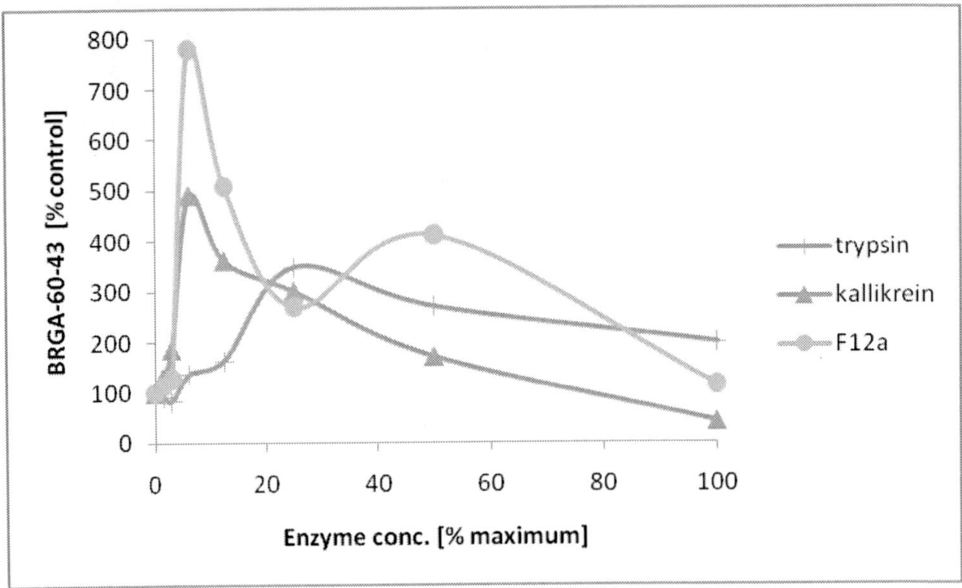

Figure 21. Stimulation of blood ROS generation by trypsin, kallikrein, F12a. The BRGA-60-43 was performed with freshest citrated blood in presence of 0-112 ng/ml trypsin, 0-92 ng/ml kallikrein, or 0-148 ng/ml F12a (final conc.) as described in Methods. The approx. SC200 values were 15 ng/ml trypsin, 3 ng/ml kallikrein, 6 ng/ml F12a.

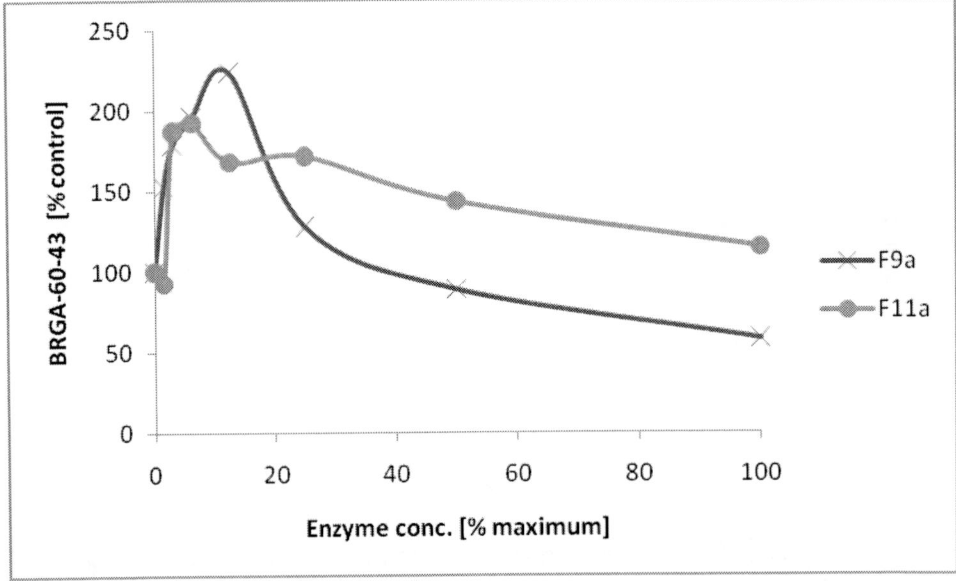

Figure 22. Stimulation of blood ROS generation by F9a or F11a. The BRGA-60-43 was performed with 0-18 mIU/ml F9a or 0-1100 ng/ml F11a (final conc.) as described in Methods. The approx. SC200 was 6 mIU/ml F9a.

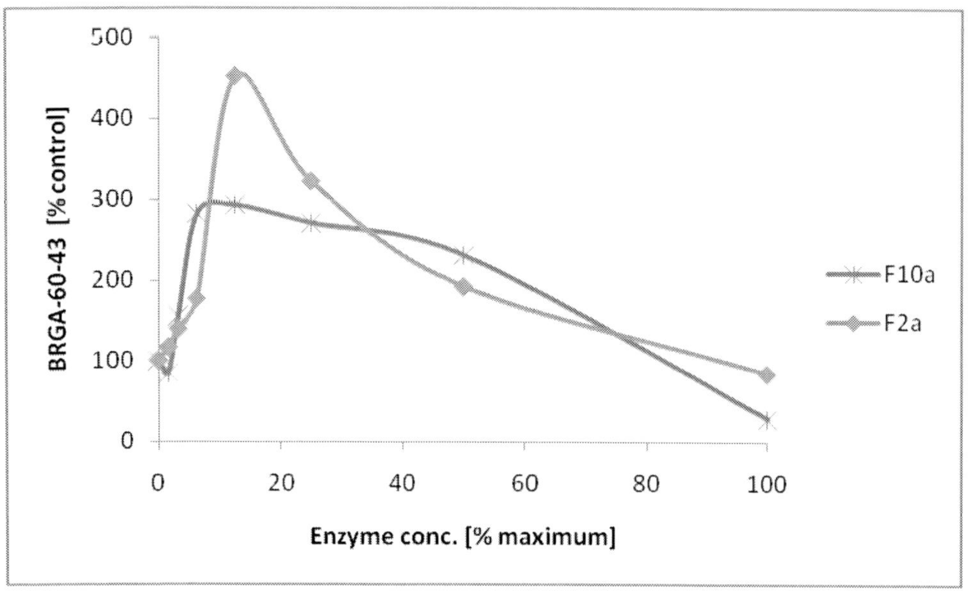

Figure 23. Stimulation of blood ROS generation by F10a or F2a. The BRGA-60-43 was performed with 0-70 mU/ml F10a or 0-6.7 mIU/ml F2a (final conc.) as described in Methods. The approx. SC200 was 3 mU/ml F10a or 0.5 mIU/ml F2a.

REFERENCES

[1]　Stief TW. Kallikrein activates prothrombin. *Clin Appl Thrombosis/Hemostasis* 2008: 14: 97-8.

[2]　Stief TW. Kallikrein triggers thrombin generation. *Hemostasis Laboratory* 2009; 2: 45-56.

[3]　Stief TW. Zn^{2+}, hexane, or glucose activate factor 12 and/or prekallikrein in two purified systems. *Hemostasis Laboratory* 2011; 4: 409-26.

[4]　Stief TW. Kallikrein activates factor 10. *Hemostasis Laboratory* 2012; 5: 211-8.

[5]　Stief TW, Klingmüller V. Diagnostic ultrasound activates pure prekallikrein. *Blood Coagulation and Fibrinolysis* 2012; 23: 781-3.

[6]　Stief T. Hemostasis activation in inflammation: kallikrein, thrombin, $^1\Delta O_2^*$/hv. In: *Thrombin and Singlet Oxygen ($^1\Delta O_2^*$) Main Factors of Hemostasis*. Stief T, ed.; Nova Science Publishers, New York, 2013

[7]　Stief TW. Coumarins trigger altered matrix (AM) coagulation activation. *Hemost Lab.* 2013; 6: 121-8.

[8]　Wachtfogel YT, Kettner C, Hack CE, Nuijens JH, Reilly TM, Knabb RM, Kucich U, Niewiarowski S, Edmunds LH Jr, Colman RW. Thrombin and human plasma kallikrein inhibition during simulated extracorporeal circulation block platelet and neutrophil activation. *Thromb Haemost.* 1998; 80: 686-91.

[9]　Stief TW. Neutrophil granulocytes in hemostasis. *Hemostasis Laboratory* 2008; 1: 269-89.

[10]　Nathan C. Neutrophils and immunity: challenges and opportunities. *Nat Rev Immunol.* 2006; 6: 173-82.

[11] Stief TW. The blood fibrinolysis / deep-sea analogy: a hypothesis on the cell signals singlet oxygen/photons as natural antithrombotics. *Thromb Res* 2000; 99: 1-20.

[12] Stief TW, Fareed J. The antithrombotic factor singlet oxygen/light ($^1O_2$/hν). *Clin Appl Thrombosis/Hemostasis* 2000; 6: 22-30.

[13] Stief TW, Fu K, Yang LH, Ramaswamy A, Fareed J. Singlet oxygen ($^1O_2$) induces selective induces thrombolysis in vivo by massive granulocyte infiltration into the thrombus. 43. GTH Congress, Mannheim, 24.-27.2.1999, *Ann. Hematol.* 1999; 78: A32 (FV112).

[14] Stief TW. The physiology and pharmacology of singlet oxygen. *Med Hypoth* 2003; 60: 567-72.

[15] Stief TW. Regulation of hemostasis by singlet oxygen ($^1\Delta O_2$). *Curr Vasc Pharmacol* 2004; 2: 357-62.

[16] Stief T. Micro-thrombi stimulate blood ROS generation. *Hemostasis Laboratory* 2013; 6: 315-25.

[17] Stief T. The routine blood ROS generation assay (BRGA) triggered by typical septic concentrations of zymosan A. *Hemostasis Laboratory* 2013; 6: 89-98.

[18] Stief T. Glucose initially inhibits and later stimulates blood ROS generation. *Journal of Diabetes Mellitus* 2013, 3: 15-21.

[19] Haverkate F, Timan G. Protective effect of calcium in the plasmin degradation of fibrinogen and fibrin fragments D. *Thromb Res.* 1977; 10: 803-12.

[20] Cao D, Mizukami IF, Garni-Wagner BA, Kindzelskii AL, Todd RF 3rd, Boxer LA, Petty HR. Human urokinase-type plasminogen activator primes neutrophils for superoxide anion release. Possible roles of complement receptor type 3 and calcium. *J Immunol.* 1995;154: 1817-29.

Chapter 14

APPLIED BIOCHEMISTRY OF THE BRGA

ABSTRACT

Background: The blood ROS generation assay (BRGA) quantifies the production of reactive oxygen species (ROS) by blood neutrophils. The BRGA screens patients for hypo- or hyper-functional neutrophils, dangerous pathophysiologies for the individual because of fungal/bacterial sepsis or autoaggression, respectively. There are up to now no informations about the BRGA under routine conditions.

Material and Methods: 144 samples of EDTA-blood were analyzed in the BRGA. 10 µl sample were added in duplicate to 150 µl Hanks' Buffered Salt Solution (HBSS modified without phenol red) in black polystyrene F-well microtiter plates (Brand®781608). 10 µl 5 mM (0.28 mM final conc.) luminol and 10 µl 36 µg/ml (2 µg/ml final conc.) zymosan A were added and the plates were incubated at 37°C. The light emissions were determined by a photons-multiplying microtiter plate luminometer (LUmo). Of particular interest is the respective maximum of each ROS generation curve and the time to reach the maximum (t-max).

Results and Discussion: There was no correlation between t-max (73±21 min) and the maximal ROS generation (5158±3924 RLU/s), r= -0.152. The blood ROS generation in BRGA-20 (about 20 min reaction time) was 78±122 RLU/s, in BRGA-40 (about 40 min reaction time) it was 1104±1380 RLU/s, r=0.568. Of particular interest is the ROS generation at about 50% of the time to reach the maximum (0.5t-max; here BRGA-40). BRGA-40 correlates with r=-0.430 with t-max. At 0.5t-max about one third of the unselected routine samples were pathological, half of them hypo- (< 100 RLU/s) and half of them hyper-active (> 2000 RLU/s). It is suggested to screen all patients for the function of their neutrophil NADPH-oxidase. The BRGA closes a gap in routine analytic.

Keywords: Reactive oxygen species, ROS, singlet oxygen, hydrogen peroxide, blood ROS generation assay, BRGA, neutrophils, fungi, bacteria, autoaggression

INTRODUCTION

The neutrophils (PMN) are our main defense cells against fungi or bacteria [1]. Severely hypo-activated PMN permit life-threatening fungal or bacterial sepsis. By contrast, hyper-activated PMN can harm the individual patient by aggressively attacking the bodie's healthy tissue. Reactive oxygen species (ROS) of the type of hydrogen peroxide (HO··OH) (original

ROS generated by NADPH-oxidase) or singlet oxygen ($^1\Delta O_2^*$) (ROS derived from H_2O_2) are the desired ROS in PMN activation. There is clinical need to screen patients for blood ROS generation [2,3].

MATERIAL AND METHODS

144 samples of EDTA-blood (1.6 mg K_3-EDTA/ml blood in polypropylene monovettes from Sarstedt, Nümbrecht, Germany) were analyzed in the BRGA. 10 µl sample were added in duplicate to 150 µl Hanks' Buffered Salt Solution (HBSS modified without phenol red; Sigma, Deisenhofen, Germany) in black high quality polystyrene F-well microtiter plates (Brand, Wertheim, Germany; article nr. 781608). 10 µl 5 mM (0.28 mM final conc.) luminol sodium salt (Sigma) and 10 µl 36 µg/ml (2 µg/ml final conc.) zymosan A (Sigma) were added and the plates were incubated at 37°C. The light emissions were measured by a photons-multiplying microtiter plate luminometer (LUmo; autobio-anthos, Krefeld, Germany) with an integration time of 0.5s/well. The respective maximum of each curve and the time to reach the maximum were determined.

RESULTS AND DISCUSSION

Figures 1-18 demonstrate the reaction kinetic in 144 unselected samples. There appeared no correlation between time to maximum (73±21 min) and maximal ROS generation (5158±3924 RLU/s), r= -0.152 (Figure 19).

The blood ROS generation in BRGA-20 (about 20 min reaction time) was 78±122 RLU/s, that in BRGA-40 (about 40 min reaction time) was 1104±1380 RLU/s, r=0.568 (Figure 20).

The distribution of the BRGA-40 activity among the 144 samples is demonstrated in figure 21. About one third of the unselected routine samples were pathological, half of them hypo- (< 100 RLU/s) and half of them hyper-active (> 2000 RLU/s).

BRGA-40 (1104±1380 RLU/s; MV±1SD) correlated with the respective maximum (5158±3924 RLU/s) with r=0.573 (Figure 22).

The time to maximum (73±21 min) correlated with the BRGA activity at 0.5fold the time to maximum (0.5t-max) (1104±1380 RLU/s) with r=-0.430 (Figure 23).

Of particular interest is the ROS generation at about 50% of the time to reach the maximum (0.5t-max). Here 0.5t-max was obtained approximately in BRGA-40. In rare cases it could be necessary to determine 0.25t-max (here approximately BRGA-20), especially if there are two ROS maxima. Only about 10% of the pathologically altered BRGA-40 values were clinically explainable (severe neutropenia [4] with BRGA ↓ or fungal sepsis with BRGA ↑ [5]). Thus, it is advisable to screen all patients for their function of the neutrophil NADPH-oxidase. The BRGA closes a gap in routine analytic.

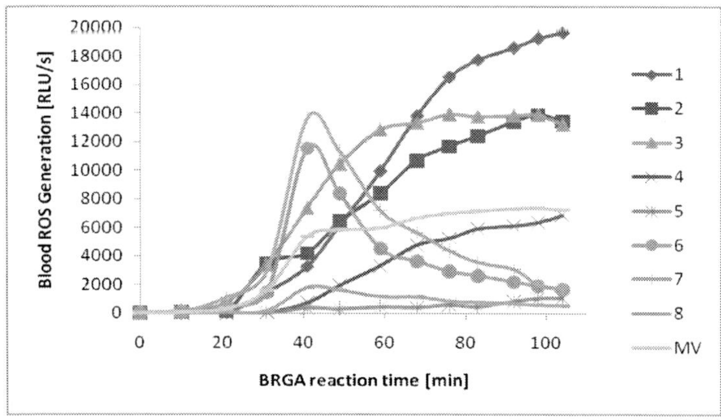

Figure 1.

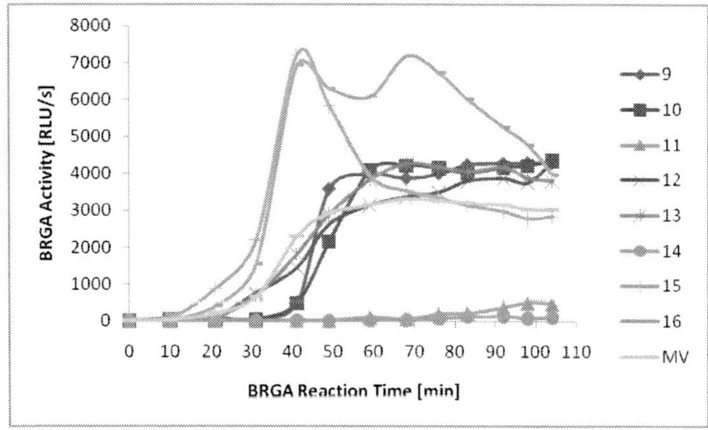

Figure 2.

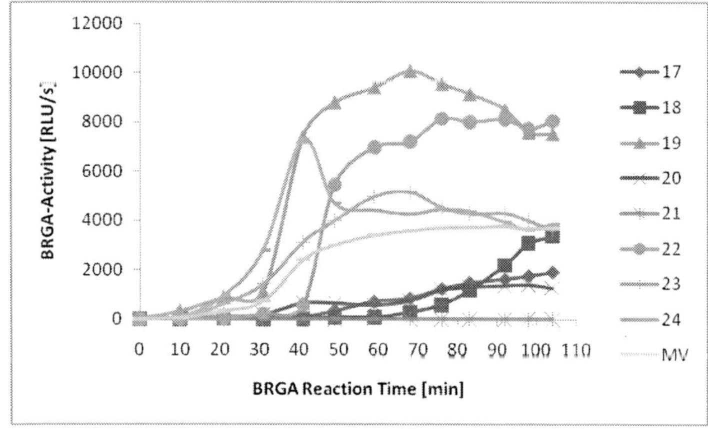

Figure 3.

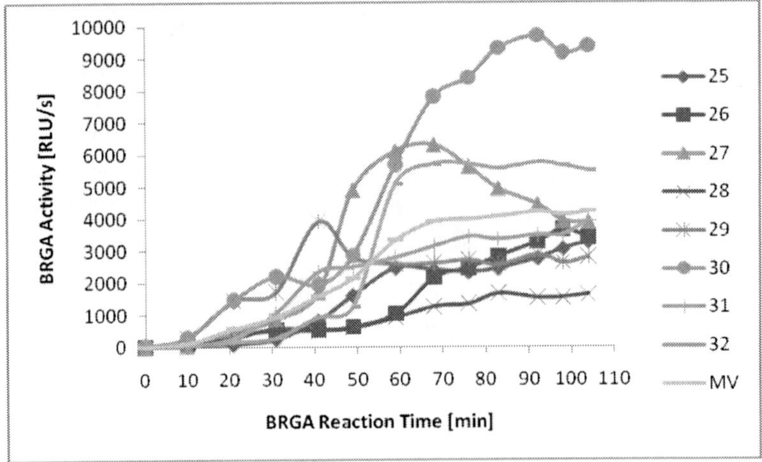

Figure 4.

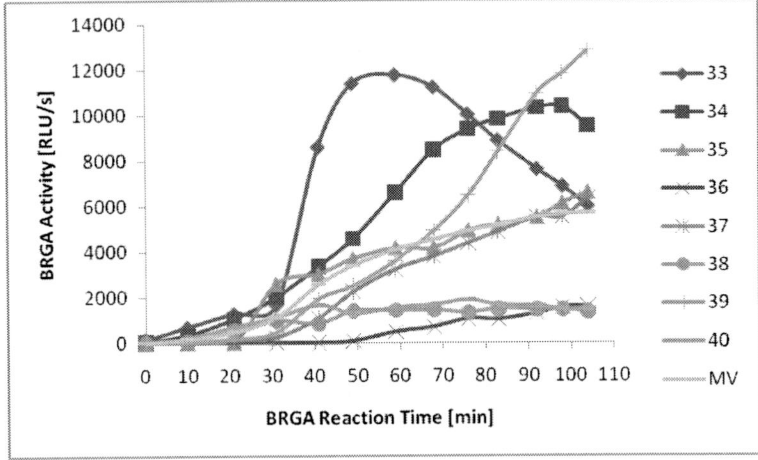

Figure 5.

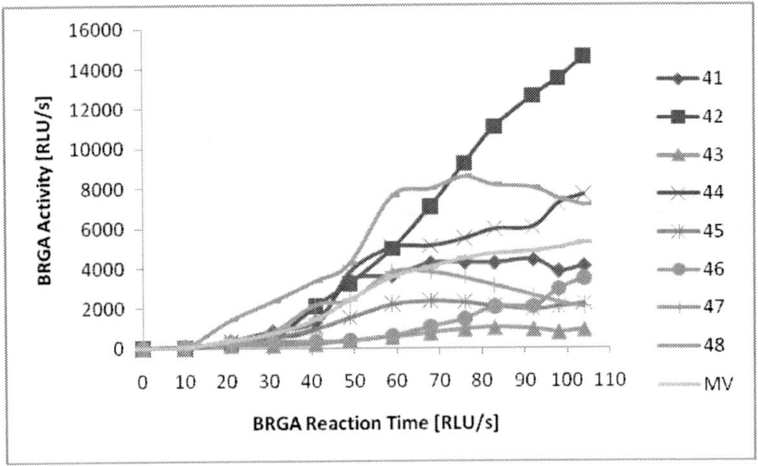

Figure 6.

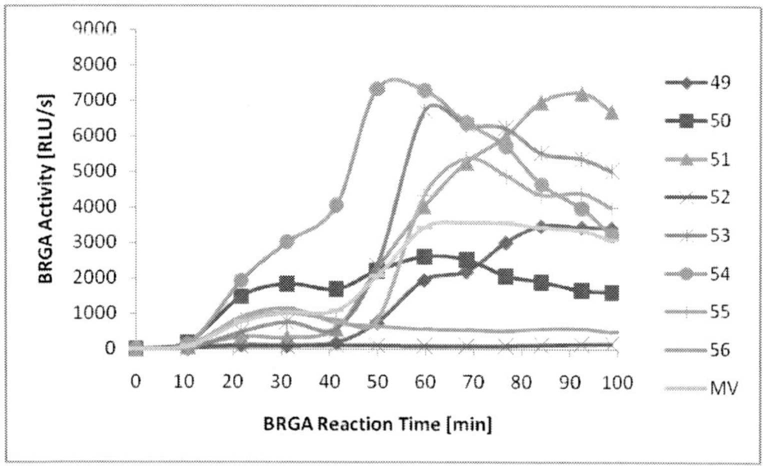

Figure 7.

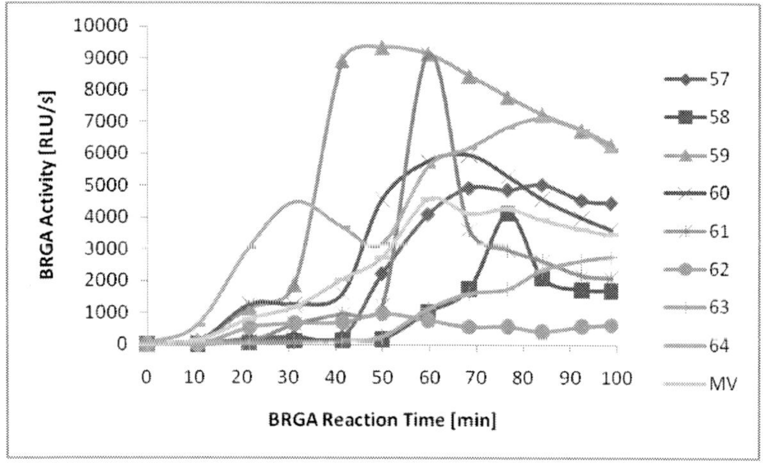

Figure 8.

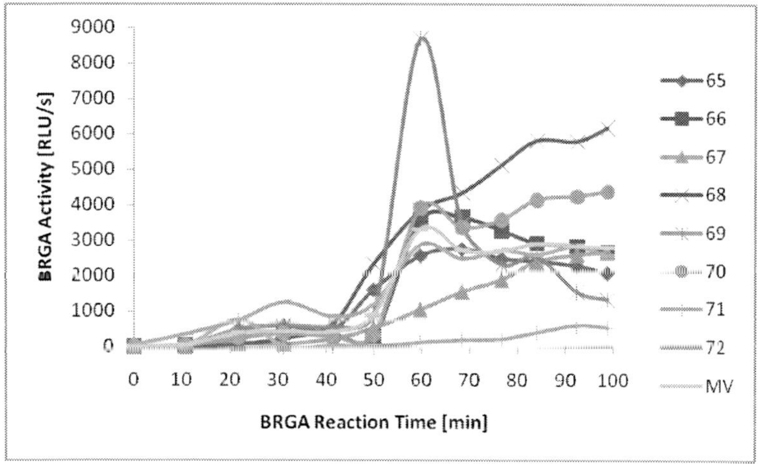

Figure 9.

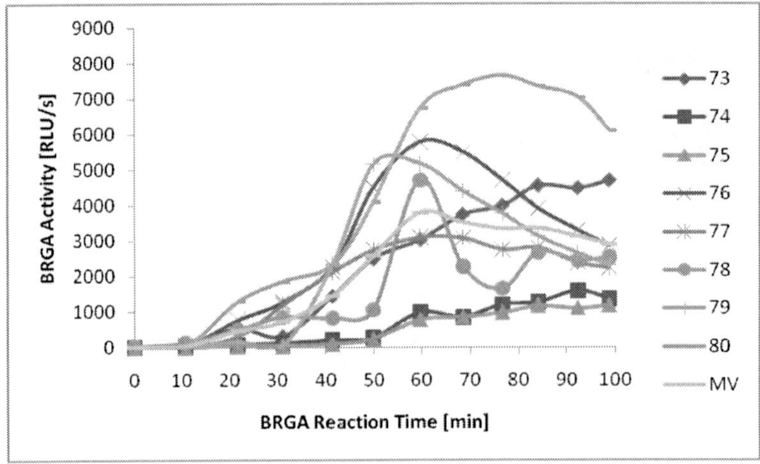

Figure 10.

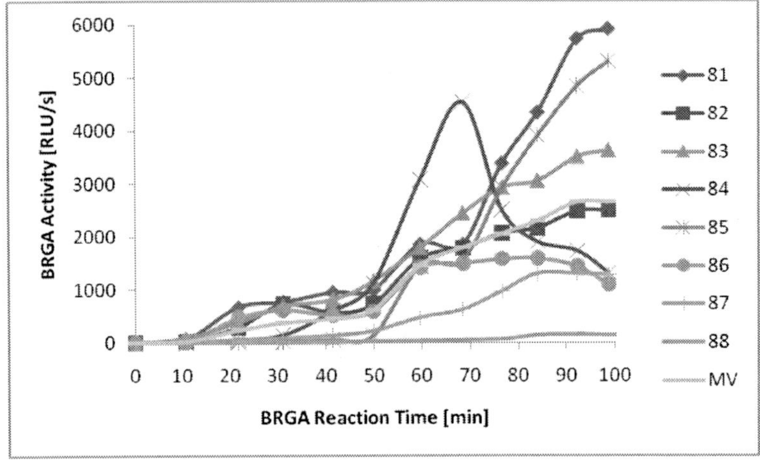

Figure 11.

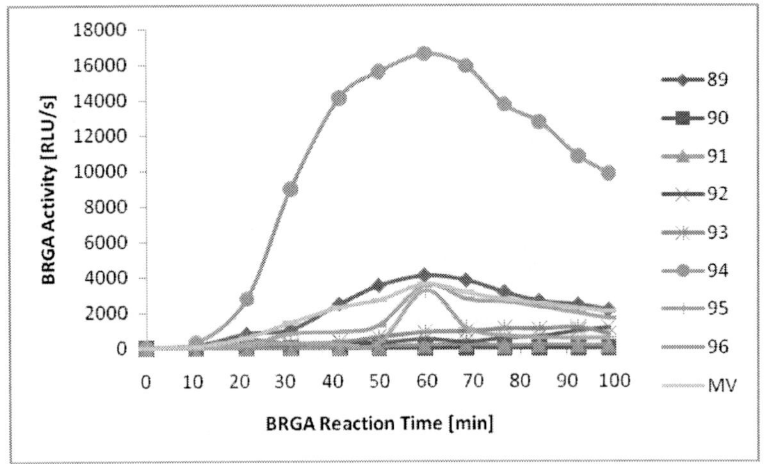

Figure 12.

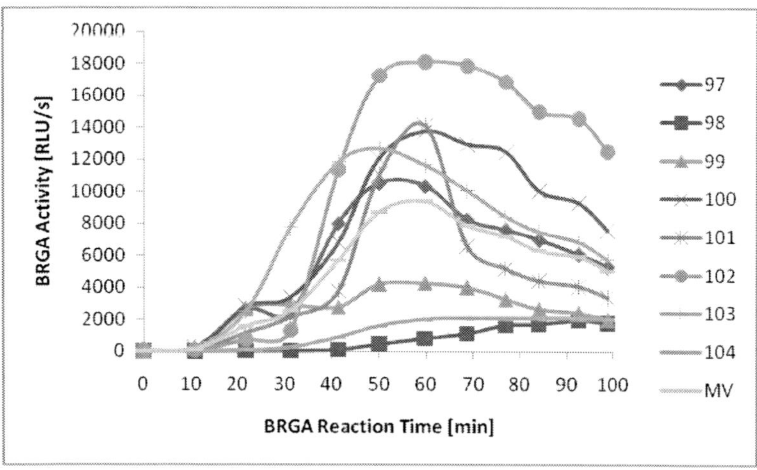

Figure 13.

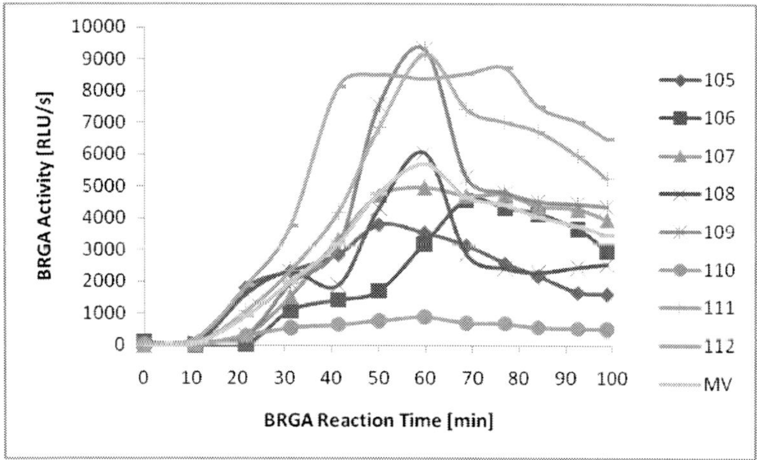

Figure 14.

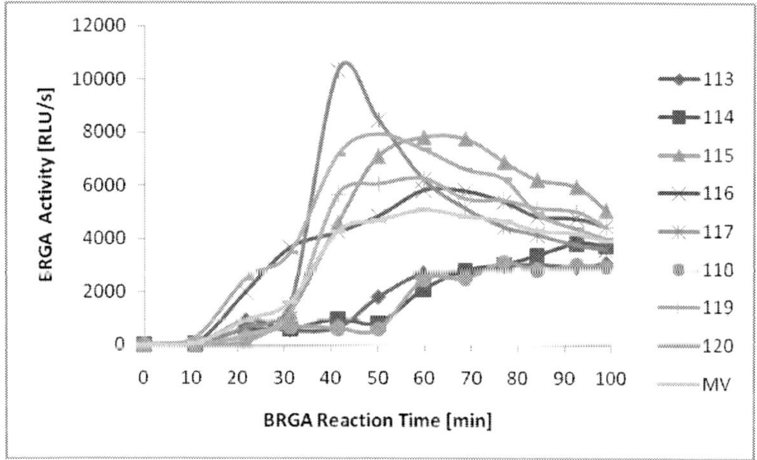

Figure 15.

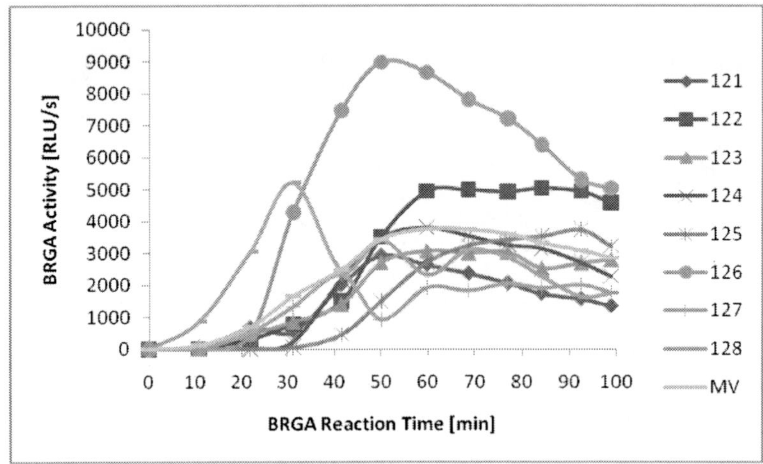

Figure 16.

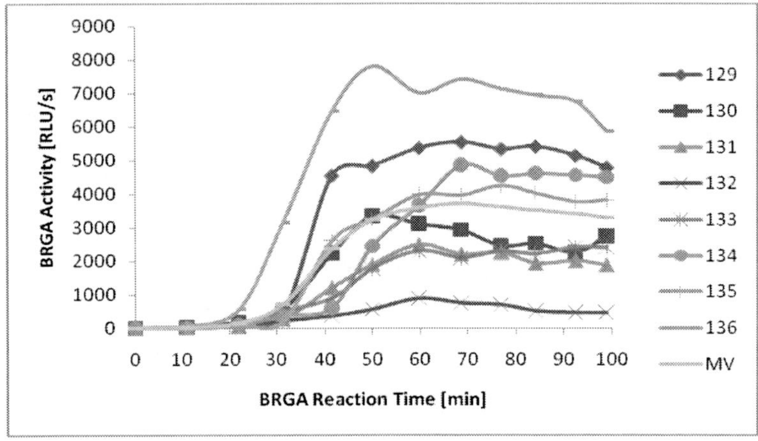

Figure 17.

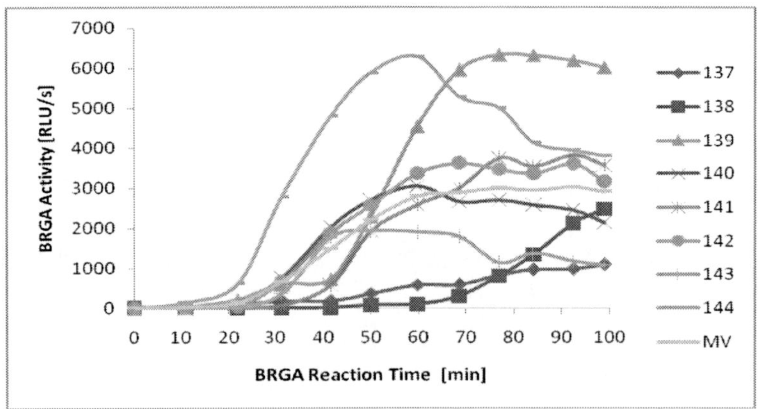

Figure 18.

Figures 1-18. BRGA in EDTA-blood. The BRGA was performed in 144 unselected EDTA-blood samples as described under Methods.

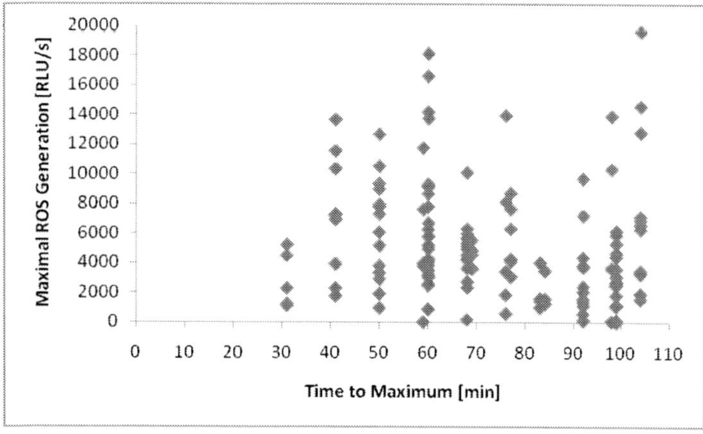

Figure 19. No correlation between time to maximum and maximal ROS generation. The 144 samples of figures 1-18 were analyzed for time to reach the maximum (73±21 min) and the amount of ROS generation at the maximum (5158±3924 RLU/s), r= -0.152.

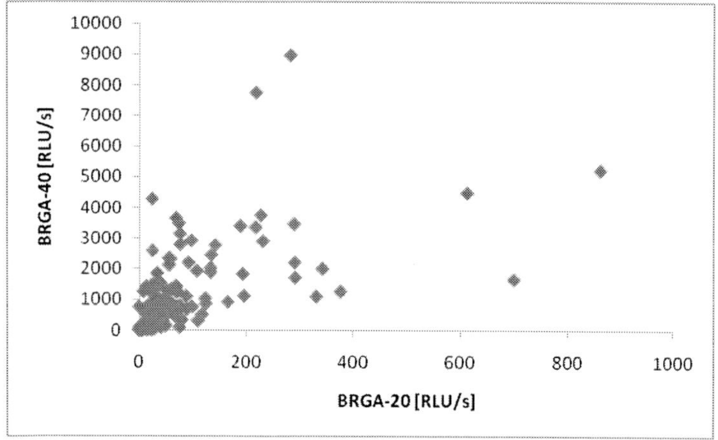

Figure 20. Correlation between BRGA-20 and BRGA-40. The blood ROS generation in BRGA-20 (about 20 min reaction time) was 78±122 RLU/s, that in BRGA-40 (about 40 min reaction time) was 1104±1380 RLU/s, r=0.568. BRGA-40 is 23±27fold higher than BRGA-20.

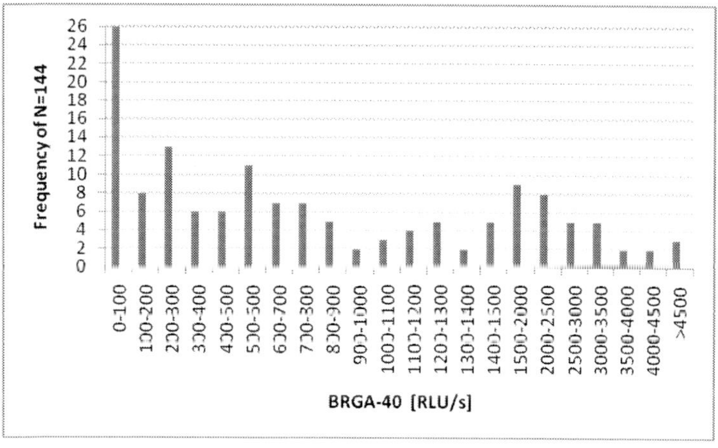

Figure 21a. Distribution of BRGA 40 activity (N=144).

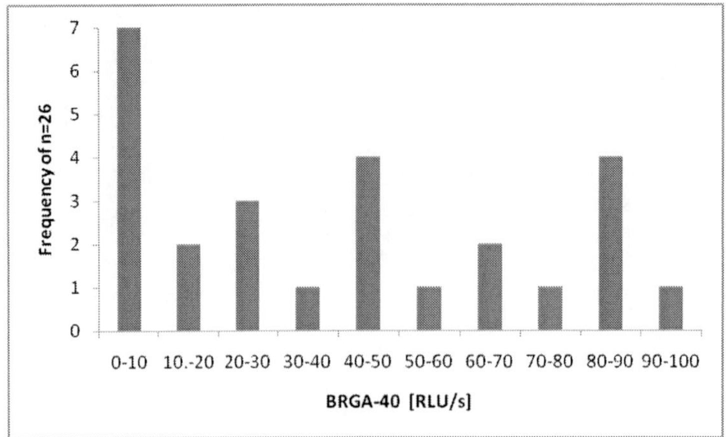

Figure 21b. Distribution of BRGA-40 activity in severely hypo-functional neutrophils (n=26).

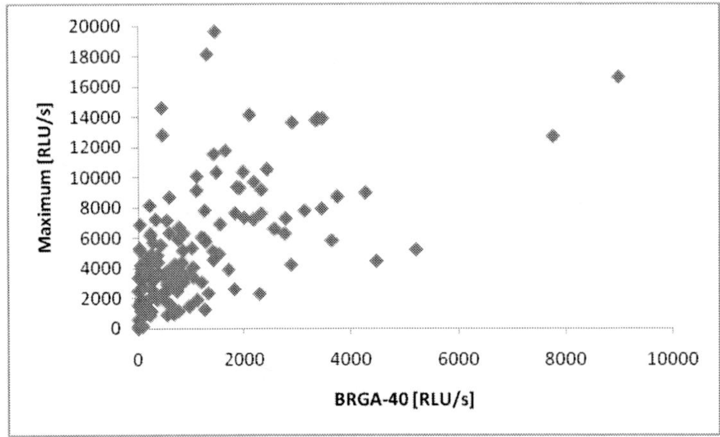

Figure 22. Correlation between BRGA-0.5tmax and maximum.
BRGA-40 (1104±1380 RLU/s) correlated with the maximum (5158±3924 RLU/s) with r=0.573.

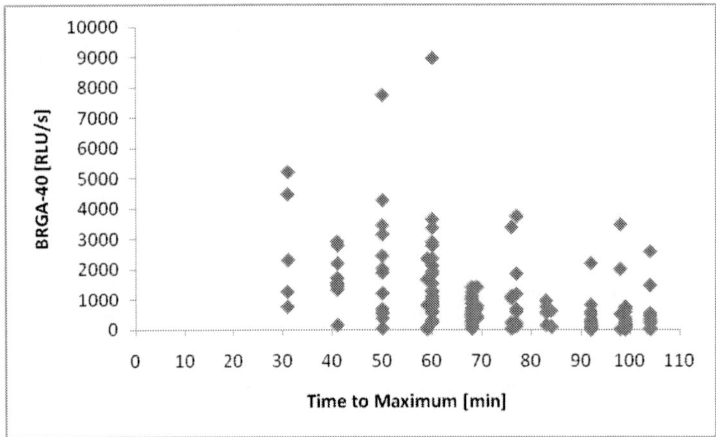

Figure 23. Correlation between time to maximum and BRGA-0.5tmax (here BRGA-40). The time to maximum (73±21 min) correlated with the BRGA activity at 0.5fold the time to maximum (1104±1380 RLU/s) with r=0.430.

REFERENCES

[1]   Stief TW. Neutrophil granulocytes in hemostasis. *Hemostasis Laboratory* 2008; 1: 269-89.

[2]   Stief T. The routine blood ROS generation assay (BRGA) triggered by typical septic concentrations of zymosan A. *Hemostasis Laboratory* 2013; 6: 89-98.

[3]   Stief T. Glucose initially inhibits and later stimulates blood ROS generation. *Journal of Diabetes Mellitus* 2013, 3: 15-21. doi: 10.4236/jdm.2013.31003.

[4]   Strydom N, Rankin SM. Regulation of circulating neutrophil numbers under homeostasis and in disease. *J Innate Immun.* 2013 Apr 5: 304-314.

[5]   Boyle KB, Stephens LR, Hawkins PT. Activation of the neutrophil NADPH oxidase by Aspergillus fumigatus. *Ann N Y Acad Sci.* 2012; 1273: 68-73.

Chapter 15

NEUTROPHILS IN FIBRINOLYSIS

ABSTRACT

The antithrombotic action of activated polymorphonuclear neutrophil granulocytes (PMN) are reviewed in this book preface. PMN with about 4000 cells/μl are the main leukocyte type in normal human blood. They are the first line of defence against invading fungi or bacteria. About 50 % of all blood PMN are slightly activated and express adhesion molecules (e.g. CD11b/CD18 = Mac-1) that bind endothelial surface proteins (e.g. ICAM-1). PMN roll over the endothelium to patrol for the existence of any pathogen that may stick to the endothelium and that may threaten vessel patency. In this sense also fibrin might be considered as a special type of pathogen. CD11b/CD18 receptors of PMN bind to fibrin and activate the cell. Activated PMN via membrane assembly of NADPH-oxidase (Nox2) produce high concentrations of H_2O_2 that generates HOCl via released myeloperoxidase. Hypochlorite reacts with amino groups (e.g. of taurine) to chloramines, an important biological source of the excited nonradical excited oxidant singlet molecular oxygen ($^1O_2^*$). $^1O_2^*$ is an inter- and intra-cellular communication signal and the selectively destructive "bullet". Light quants (hv, especially those at about 300-400 nm) enhance PMN patrolling and direct and recruit PMN to a pathologic focus, such as a pathologic thrombus. CD11b/CD18 binding induces H_2O_2 generation and viceversa.

$^1O_2^*$ is a selective oxidant. Due to its nonradical nature, $^1O_2^*$ is relatively inoffensive to normal mammalian cells but highly destructive against pathogens. $^1O_2^*$ inactivates important components of hemostasis, such as thrombocytes, fibrinogen, factors 5, 8, 10, α2-antiplasmin, and PAI-1 with 50 % inhibitory concentrations of 1-2 mM in blood or plasma. It is suggested to imitate the physiologic PMN mediated thrombolysis by intravascular injection of violet photons or of sub-millimolar concentrations of chloramines.

Keywords: Neutrophils, polymorphonuclear granulocytes (PMN), hemostasis, fibrinolysis

INTRODUCTION

Polymorphonuclear neutrophil granulocytes (PMN) with a diameter of about 12-15 μm have a poly-segmented lengthy nucleus, many cytoplasmic granules, large glycogen stores, and a highly convoluted surface [1]. PMN as cells of the innate immune response guarantee sterility of blood or of alveoles [2-8]. PMN are with about 4000 ± 2000 cells / µl the main leukocyte type in flowing blood. These flowing cells are about 50 % of all blood PMN, the other 50 % of all blood PMN are rolling over the endothelium, i.e. normally 50 % roll and 50 % flow [9]. Flowing PMN are inactive, rolling PMN are slightly activated. Rolling PMN patrol for invading pathogens (e.g. fungi or bacteria) that are attached to the endothelium. Pathogens strongly activate PMN, activated PMN fight against these pathogens [10], the PMN release pro-inflammatory cytokins such as interferon-gamma (IFN-γ) [11], interleukin-1 (IL-1), tumor necrosis factor–alpha (TNF-α), or granulocyte-colony stimulating factor (G-CSF) [1,12-14]. Pathological deposits of fibrin on endothelial cells (pathological thrombi) threaten vessel patency, and might also be considered as a type of pathogen. The present review focuses on the interaction of PMN with fibrin.

PMN Rolling and Diapedesis

The half-life of PMN in the blood circulation is only about 9h [1]. The physiologic PMN rolling over the endothelium is primarily mediated by reversible binding of PMN CD11b/CD18 to endothelial intercellular adhesion molecule (ICAM) - 1 [15,16], only secondarily by selectin/selectin receptor interactions, e.g. binding of endothelial selectins to PMN selectin receptors [17-19].

CD11b/CD18 of slightly activated (primed [20]) PMN not only bind ICAM-1 but also fibrin [21,22]. Enhanced CD11b/CD18 binding results in increased cell activation and viceversa [23,24,195,196]. Priming means preparing for strong activation (triggering; e.g. by zymosan A). Singlet oxygen is a typical primer, organizing the cytoskeleton for the following assembly of Nox2 [236]. Rac GTPases participate in cytoskeleton building [238].

PMN respond to cell activators such as plasmatic lipopolysaccharide (LPS) at about 3-10 ng/ml with enhanced adhesion [1, 23]. LPS concentrations > 100 ng/ml are toxic for PMN [1, 24, 25] resulting in defective oxidant production [23]. Thus, there is great clinical need to examine the plasmatic reactivity of endotoxin (free endotoxin) in relation to its pathophysiology [23]. The monocytes/macrophages seem to play a much greater role in the pathophysiology of sepsis and pathologic intravascular coagulation (PIC) than the PMN [26,153,154]: only about 5 ng/ml LPS added to normal citrated blood induced a destruction of about 20 % of blood monocytes within 60 min (37°C), the monocytes count decreased linearly as a function of LPS incubation time; the PMN were resistant to up to 100 ng/ml LPS in blood [154]. Monocyte derived microparticles trigger intrinsic hemostasis [24].

Glucocorticoids reduce the adhesion of PMN to the endothelial surface [27,35], possibly by downregulation of the expression of CD11b, CD18, or L-selectin (CD62L) [28-31] by decreasing chemotaxis [32,33], or by proteolytic cleavage and release of L-selectin (CD62L) or CD18 from the surface of the PMN [34]. 1h after injection of 200 IU ACTH there appeared an about 10fold increase in plasmatic corticoid concentration (basal conc. = 3 ng/ml); 2h after injection of ACTH the blood PMN (not other blood leukocytes) began to increase, 4h after ACTH injection the PMN count approximately doubled [1].

The nucleus of the PMN is multi-segmented and lengthy. So it allows the PMN to line up its nucleus in a straight line, which is a prerequisite for rapid migration through the endothelium [1]. To reach a pathogenic focus in the interstitium, blood PMN traverse the endothelium [36] in a process that does not involve proteolytic destruction, and PMN traverse the basal lamina in a process that may involve proteolysis [37-39]. *In vivo*, PMN mostly penetrate endothelial cells through small openings called *"fenestrae"* of the cell membrane [36, 37, 40, 41]. These pores of about 50 nm in diameter, whose normal function might be to enable plasma contact with the subendothelial tissue, can widen up to a size of 3-5 μm, if a PMN penetrates the endothelium [36, 42]. The PMN looks for thin endothelial regions, which could be opened *fenestrae*, or zones where the upper and the lower membrane of the endothelial cell are in close contact [36]. During and after penetration the cellular cytoskeleton is rearranged [43]. *In vitro*, PMN might penetrate the endothelium also by paracellular lysis of tight junctions [36].

PMN GRANULES

Neutrophils possess four different types of granules that were classified according to the moment they occur in the maturing cell [4]:

1. primary granules
2. secondary granules
3. tertiary granules
4. secretory vesicles.

The primary granules contain myeloperoxidase, proteases (e.g. urokinase, elastase, cathepsins), phospholipases (PLA_2) [44], and polycationic peptides/proteins (e.g. defensins). Myeloperoxidase is the enzyme that generates the powerful oxidant HOCl out of H_2O_2 and Cl^-, the proteases can degrade peri-cellular proteins. The polycations are toxic for germs and enhance the activity of the proteases [45,46] and activate PMN [4].

The secondary granules are also known as the specific granules; they contain the cytochrome b558 components gp91phox and p22phox, alkaline phosphatase, lactoferrin and gelatinase B (= matrix metalloprotease 9 = MMP9) [4]. Tertiary granules are similar as secondary granules, but they are bigger than primary or secondary ones and contain especially lactoferrin and polycations [1]. The secretory vesicles contain cytochrome b558, alkaline phosphatase, CD16, decay accelerating factor, and albumin as stabilizer [47].

The granules are translocated to the plasma membrane (or to the membrane of a phagosome) after assembly of microtubules and remodelling of the actin cytoskeleton [4,48]. In the translocation machinery with a possible GTP-ase switch, actin is first polymerized to "steel-ropes" (flexible cytoskeleton [180]), along them the granules are then transported to the plasma membrane, and myosin may act as "motor" [4,47]. Secretory vesicles are first released, then the tertiary / secondary granules. The primary granules that contain myeloperoxidase are the last ones to be released, i.e. myeloperoxidase is released not until H_2O_2 production has started. The degranulation is regulated by phospholipid mediators (e.g. phosphatidyl-inositol-4,5-diphosphate (PIP_2) [4].

OXIDANT METABOLISM

PMN generate high concentrations of reactive oxygen species (ROS) by a combined action of membranous NADPH-oxidase and released myeloperoxidase. The main reactive ROS are hydrogen peroxide (H_2O_2), hydroxyl radical ($\cdot OH$), and excited singlet molecular oxygen ($^1O_2*$). First, the NADPH-oxidase generates $\cdot O_2^-$ that immediately dismutes to H_2O_2 [168,238] that may break to 2 $\cdot OH$, or that may react to HOCl (hypochlorite) by myeloperoxidase catalysis. HOCl reacts with amino groups to chloramines. Chloramines or HOCl can generate $^1O_2*$ by reaction with H_2O_2; $^1O_2*$ reacts with C=C bonds to yield excited carbonyls (R-C=O*) that emit at 300-400 nm (depending on R) or with sulphur atoms in methionine or cysteine [85]. The ROS type might be modulated by certain cellular enzymes, cofactors, or reaction conditions [165-167]. The neutrophil is the main source of ROS in the human organism, the cell disposes of an enzyme system whose principal function is to produce ROS. Activated neutrophils consume 20fold more oxygen utilizing glucose to generate NADPH by the pentose phosphate pathway [234]. NADPH is metabolized by Nox2 at K_m values of only 40-45 µM [238].

There are other cellular systems that generate ROS as a byproduct: the mitochondrial respiratory chain, peroxisome oxidases, cyclooxygenases, lipoxygenases, uncoupled NO-synthases, uncoupled CYP450 [210], CYP 450 reductases, xanthine oxidase [234,236].

ROS at high concentrations act as bullets, at low concentrations they may also act as cell signals. Cellular signalling is mediated by inhibition of phosphatases, activation of kinases, regulation of ion channels (membrane K^+ channels; membrane or intracellular Ca^{2+} channels), or Ca^{2+} pumps. The S-glutathiolation Ca^{2+} ATPase pump is activated by low concentrations of ROS, high concentrations inactivate it [238].

Especially $^1O_2*$ might play an important role in inter- and intra- cellular communication. PMN might possess membraneous (opsin-like or chromophore-like [207,208,213]) receptors for singlet oxygen and/or photons, e.g. the chemotactic IL-8 receptor belongs to the rhodopsin-like serpentines that are receptors with 7 transmembrane helices and that are coupled to G-proteins [155]. Retinoids (vitamin A and its derivatives) are involved in a broad range of events of cell maturation, differentiation, and signalling that are up to now not completely understood [157,158,159,161]. Interestingly, all-trans retinoic acid (ATRA) is an important regulator of granulocyte differentiation [49,160]: ATRA up-regulates the expression of CD11b, CD11c, CD15, CD65, CD54, CD38. ATRA establishes complete remissions in acute promyelocytic leukaemia (M_3), a feared hyper-acute form of leukaemia. Especially the pathologic intravascular coagulation or the primary hyperfibrinolysis, frequent complications of M_3, are avoided by oral ATRA prior to definitive chemotherapy [156,161].

NADPH-OXIDASE

Reduced nicotine amide adenine dinucleotide phosphate (NADPH) is the intracellular cofactor of an important PMN weapon: the NADPH-oxidase. NADPH is generated by cellular uptake of glucose [52,53], degradation of cellular glycogen stores, phosphorylation of glucose, and action of the glucose-6-phosphat-dehydrogenase. The NADPH-oxidase is a complex of proteins that are membrane bound or that originally belong to the cytosol of the cell. The cytosolic subunits of the NADPH-oxidase are the Nox activator p67phox, that with its NADPH binding site is of importance for normal oxidase function [54], p47phox [55], p40phox, which is of minor importance [47], and the p91phox regulatory cytoplasmic G-protein Rac (Rac1 or Rac2) recruited by p67phox to the membrane [56, 235]. For activation of NADPH-oxidase, p47phox (the Nox organizer [232]) must be serine-phosphorylated (the limiting step in Nox2 assembly) to bind p22phox [233,235,236,238] and p40phox has to be separated from p67phox, i.e. initially p40phox inhibits NADPH-oxidase assembly, later is may stimulate Nox2 activity [57,238]. The gene loci are Xp21.1 (gp91phox, main part of Nox2), 16q24 (p22phox), 7q11.23 (p47phox), 1q25 (p67phox), 22q1,3.1 (p40phox) [238].

Pathways linked to Nox-activation include phospholipases (PLC, PLD), arachidonic acid metabolites, protein kinase C (PKC), phosphatidylinositol 3-kinase (PI3K), GTP-binding proteins (Ras, Rac1, Rac2), chaperones (protein disulphide isomerase), transcription factors (e.g.NF-kB), MAPK family members, non-receptor protein tyrosine kinases; phosphorylation and redox cascades (NF-kB, AP-1, STAT1/3) are of primary importance in Nox activation; ROS amplify their own generation [232,236]. Antioxidants (as vitamin C) often do not inhibit active Nox because the substances are CYP450 transformed into redoxcycling H_2O_2-generators [236]. The pteridine neopterin inhibits Nox2 in low micromolar conc., but upon prolonged incubation it is (CYP450 ?) metabolized into a ROS generator [238].

There are more Nox2 – like enzymes in the human organism: Nox1 in colon, smooth muscle, endothelium, Nox3 in the inner ear, Nox4 in kidney, blood vessels, Nox5 in lymphoid tissue, testis, DUOX1 and DUOX2 in thyroid [238].

The membranous subunits of the NADPH-oxidase are gp91phox and p22phox that are the ß-chain and the α-chain, respectively, of cytochrome b558. Removal of carbohydrates (mainly N-acetyl-glucosamine and galactose) at Asn132, Asn 149, Asn 240 in the second and third extracellular loop of gp91phox results in a 55 KDa protein, i.e gp91phox is highly glycosylated [238]. The abbreviation phox stands for phagocytic oxidase (NADPH - oxidase 2 = Nox2). 1 molecule of cytochrome b558 binds also 1 molecule of Rap1a, a 22kDa guanosine 5`-triphospate (GTP) – binding protein that is also present in the specific granules [58]. Active phospholipase C mediates Rap1a activation. Rap1a and Rac might be switches of the NADPH-oxidase [59,60]. Inactive Rac binds GDP, active Rac binds GTP [238].

About 85 % of the cytochrom b558 molecules are within the specific granules or secretory vesicles of an inactive (resting) cell, about 15 % of the cytochrome b558 molecules are already located in the plasma membrane [58]. Upon cell activation the active NADPH-oxidase is assembled at the cell membrane by translocation of the membranous Nox2 subunits and of the cytosolic Nox2 subunits to the cell membrane (or to the membrane of a phagosome). In inactive (resting) PMN the cytochrome b558 containing granules are randomly distributed all over the interior of the cell (see e.g. Figure 1A of [47]). Upon activation, actin "steel-ropes" are formed by focal polymerization that enable the transport to the plasma membrane (see e.g. Figure 1B of [47]). Finally, the complete NADPH-oxidase is assembled at the plasma membrane, the PMN looks like a "tank" with multiple "guns" directed towards all directions (see e.g. Figure 1D of [47]). High concentrations of H_2O_2 are generated pericellularly: 1 million activated PMN generate about 100 nmol H_2O_2 [61], i.e. considering areas of PMN infiltration with about 1 million activated PMN/µl, these PMN might generate a micro-ambience of H_2O_2 in the 100 mM range. The PMN myeloperoxidase transforms the rather radicalic H_2O_2 into slightly radicalic HOCl [62] that e.g. reacts with amines, e.g. taurine, to nonradicalic chloramines, mainly taurine-chloramine. Chloramines are anti-fungal, anti-bacterial, anti-viral, anti-parasitic, anti-tumoral, and anti-thrombotic [63-82, 85, 92, 213]. Thus, chloramines act against all pathogen-types. PMN might generate ambiences of about 1-10 mM chloramine [10, 83]. H_2O_2 and HOCl/chloramines might act synergistically [84]. Chloramines are important physiologic sources of the nonradical excited oxidant singlet oxygen ($^1O_2{}^*$) [85] that at neutral-physiologic pH selectively reacts with sulphurs (in methionine or cysteine) or with C=C groups [86, 87]. The latter reaction yields excited carbonyls (R-C=O*), that emit photons of about 300-400 nm (in the UVA to blue spectrum of light), which might be of physiologic importance for specific cell signalling within the dark blood stream that is partially filled with red cells [88,233]. $^1O_2{}^*$ is anticoagulant, profibrinolytic, and complement activating [89-94]. Nonradical oxidants are well tolerated by normal mammalian cells [95,96].

gp91phox is the most important part of the NADPH-oxidase: intra-cellularly it binds the NADPH binding protein p67phox, intra-membranously it contains FAD (flavin adenine dinucleotide) and 2 hemes (one inner, one outer), extracellularly it contains a binding site for molecular oxygen (O_2) [56]. So, gp91phox is the main transporter of electrons from inside to outside the cell, i.e. the gp91phox part of the NADPH-oxidase is responsible for the electrons – flop (Figure 1; movement from inside to outside = flop, movement from outside to inside = flip).

$^1\Delta O_2{}^*$

Taurine ↑

OCl⁻

MPO ↑

Outside cell

$O_2 \rightarrow H_2O_2$

Membrane **p22phox** <u>**gp91phox FAD**</u> $\Rightarrow$ **inner Heme** $\Rightarrow$ **outer Heme** ⇑ ⋮ ⋮ **Rac1** ⇑

Inside cell **p47phox** <u>**p67phox NADPH**</u> **+ H⁺** $\rightarrow$ **NADP⁺**

Figure 1. ROS generation by assembled NADPH-oxidase and released myeloperoxidase Electrons from inside the cell (in blue), bound in C-terminal NADPH of p67phox, are transported to FAD and the inner and outer heme of gp91phox and to molecular oxygen (O2) outside the cell. gp91 phox is the main intramembraneous component of the Nox2 [8,27]; gp91phox consists of 6 transmembrane domains, 3 extracellular loops, the two histidine-fixated hemes (◊) His101··◊··His209 and His 115··◊··His222 [238]. The GTPase Rac1 is the switch. The electrons are transferred to O2 outside the cell or inside a phagosome forming ·O2- that instantaneously dismutes to the stable basal reactive oxygen species (ROS) H2O2. Secreted myeloperoxidase (MPO) generates hypochlorite that reacts especially with taurine to a chloramine, the N-chlor-taurine (tau-Cl) [64,65]. Tau-Cl oxidizes via the selective ROS singlet oxygen (1ΔO2*). The flux of electrons is indicated by the symbol ⇑. The activity of the proton channel in the 3rd histidine-rich transmembrane domain of gp91phox is directly proportional to Nox2 activity [238].

The activity of the NADPH-oxidase is regulated by protein phosphorylation [50,58], e.g. by protein kinase C, furthermore alkaline phosphatase for phosphorylation of membrane proteins [97] is associated with gp91phox and p22phox within the same cell granules [47].

Three isoforms of NADPH-oxidase are expressed in the vascular wall: endothelial cells and adventitial fibroblasts have a NADPH-oxidase similar to that of PMN with gp91phox = Nox2 as the main component [98]; the NADPH-oxidase of vascular smooth muscle cells is different to gp91phox, these isoforms are named Nox1 and Nox4 [99,100].

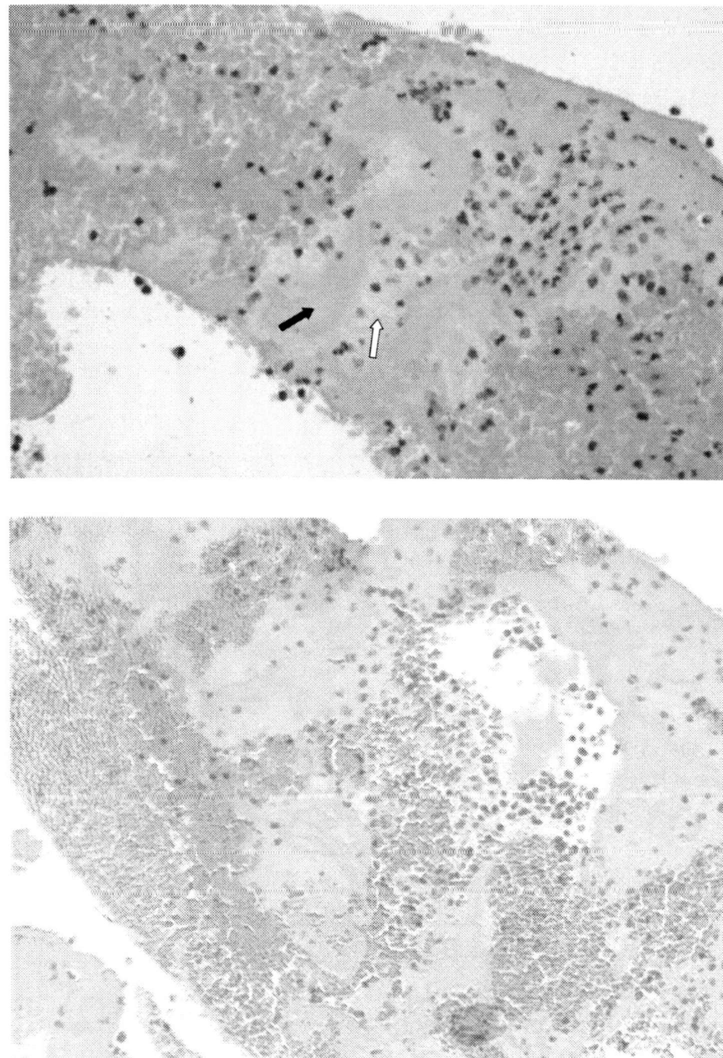

Figure 2. Thrombolysis *in vivo* by PMN. Thrombi were generated in jugular veins of rabbits by injection of 0.2 U/ml FEIBA and 15 min blood stasis in this vein segment [92]. Chloramine-T® (0.5 mM; $^1O_2^*$ generator) was infused over 30 min. 60 min after drug-infusion the jugular veins were excised. The minimal thrombi (<5% of untreated controls) were histologically compared with the full control thrombi (both HE-stained). PMN massively infiltrated the thrombi, if the animal was treated with $^1O_2^*$, the PMN/erythrocytes ratio in animals without $^1O_2^*$ therapy of about 1/2000 could increase more than about 1000fold in favour of the PMN [92]. Areas of destroyed erythrocytes are coloured in dark-pink (➜), areas of pericellular fibrinolysis around single PMN with less erythrocytes, that were fibrinolytically released from the thrombus, are coloured in white to light-pink (⇒).

SINGLET OXYGEN ($^1\Delta O_2^*$) AND HEMOSTASIS

The nonradical excited reactive oxygen species (ROS) $^1O_2^*$ is a molecule of both cellular signalling and destruction. At low doses ROS behave as cell signals: they regulate cell cycle, protein kinases, and gene expression [99,101,102]. At high doses $^1O_2^*$ is pathogen-destructive and signalling: UVA to blue light quants of about 300-400 nm derived from the $^1\Delta O_2^*$ reaction product excited carbonyl (R-C=O*) are emitted, when PMN are strongly activated. In the dark blood stream they may act as inter- or intra- cellular signals. This signalling side effect might be comparable to the gun fire of soldiers against invading bandits at night: nearby other soldiers see the exact location of the battle, are recruited to this crime scene, and amplify the gun fire against the invasive bandits.

The expression of gp91phox mRNA in the artery wall is upregulated during atherogenesis [99]. Hydroxyl radicals (·OH) may also be formed, especially if transition metal catalysts (Fe^{3+}/ Fe^{2+}) are interacting (Fenton reaction) [58,99]. For normal mammalian cells ·OH is rather dangerous, it is extremely reactive and does not diffuse more than a few molecular diameters from its site of formation [236]. Therefore, ·OH unspecifically breaks protein chains or forms tyrosine dimers that might crosslink proteins [103,104]. $^1O_2$ or inhibitors of ·OH [105] are rather good. Whereas it seems to be clear that the genesis of ·OH promotes atherogenesis, the generation of subtoxic levels of the nonradical $^1O_2$ by gp91phox (=Nox2), Nox1, and Nox4 might well be anti-thrombotic or anti-atheroscerotic [106]: patients with chronic granulomatous disease (CGD), an illness of dysfunctional NADPH-oxidase (defective phox proteins gp91phox, p22phox, p67phox, p47phox [56]), often suffer from severe atherosclerosis [58,107,108].

Hyperoxia stimulates ROS production by Nox proteins, hypoxia involves the mitochondrial electron transport to generate H_2O_2 [237].

Atherosclerosis development in mice cannot be inhibited by a deficiency of PMN NADPH-oxidase [109]. In the pre-phase of atherosclerosis, that might be called thrombo-atherosclerosis, there appears a pathological fibrin clot. This clot can be lysed by recruited PMN that is activated by $^1O_2^*$ generating endothelium; PMN or endothelium can be activated by bound fibrin [110], by activated platelets, or by other activated PMN [102,111-120]. PMN via CD11b/CD18 bind to the C-gamma-terminus of D-domains in fibrin [115] and become activated. Consequently, a $^1O_2^*$ and R-C=0* (excited singlet oxygen or excited carbonyl) - mediated decrease of the count of flowing blood PMN in favour of PMN that chemotactically adhere to and infiltrate the pathologic clot on the endothelium might be physiological [121-123,162,163,164]. In contrast, in the late phases of atherosclerosis, infiltration of tissue factor (TF) expressing macrophages (MΦ) [124-128] but also of PMN [129] that via ROS might enhance TF production or PLA_2 activity of MΦ [130] might be pathological. Patients with dysfunctional PMN, e.g. patients positive for anti-cytoplasmic antibodies (ANCA) often suffer from thrombosis [131,132,178].

Also fibroblasts bind fibrin via ICAM-1 [133]. Activated fibroblasts transform the immature fibrin clot into mature fibrotic tissue, i.e. thrombosis might be a pre-form of fibrosis [134].

CELLULAR FIBRINOLYSIS

Fibrinolysis consists of a plasmatic part and a cellular one [135,136]. The plasmatic part is the plasmatic fibrinolysis system with plasmin as the central enzyme. The cellular part are the PMN with their urokinase, elastase, and singlet oxygen [51,138-140]. If pathologic amounts of fibrin are deposited over endothelium, that are beginning to obstruct the vessel lumen, the blood PMN are activated. If the PMN system is insufficient, e.g. because there is too much fibrin as e.g. in PDIC, this cellular fibrinolysis is frustrane [141]. D-dimer degradation by PMN elastase makes the D-dimer assay difficult to interprete, since the plasmatic D-dimer concentration would then depend on at least 4 totally different hemostasis enzymes: thrombin, factor XIIIa, plasmin, and elastase. Circulating basal plasmatic thrombin is much better suited for diagnosis of coagulation activation, e.g. in sepsis [142,211,212].

Corticoids suppress PMN adhesion to the endothelium and result into thrombosis [143-146]. Physiologic PMN activators such as singlet oxygen or G-CSF induce enhanced adhesion to endothelial cells via CD11b/CD18 – ICAM-1 interaction and subsequently into thrombolysis [83,147-150]. PMN recruitment to a pathologic thrombus is a prerequisite for efficient thrombolysis, mediated by singlet oxygen, urokinase, plasmin, and elastase [151,152].

Imitation of the physiologic PMN mediated thrombolysis is of major clinical interest to avoid severe bleeding complications. PMN can be activated e.g. by about 2 µg/ml G-CSF [83,179], by about 0.5 mM chloramine [92,147,213], or by usual therapeutic doses of vancomycine that precipitates fibrinogen [89,92] to a possible CD11b/CD18 activator (about 0.25 mM of vancomycine activates blood PMN comparably efficiently as about 0.5 mM of chloramine). Blood concentrations of about 0.5 mM chloramine do not significantly alter the normal blood coagulation because the 50 % inhibitory concentration of chloramine against susceptible coagulation components is about 1 mM [213].

CONCLUSION

For several decades there has been a controversial if leukocytes are the cause or the consequence of pathologic coagulation activation. Now it seems to be clear that M∅ with their TF + phospholipase A_2, that are potent activators of extrinsic and intrinsic coagulation, respectively, can be rather pathologic [197-205,214], and PMN with their $^1O_2$* + urokinase, that mediate peri-cellular fibrinolysis (see e.g. PMN as in the right upper corner of figure 4B of [172]), are rather physiologic [169-171,175-177]. A fresh thrombus is first invaded by PMN; M∅ accumulate in the thrombus if it is older than 3 days [215-218]. Blood levels of myeloperoxidase increase upon phagocyte activation, especially upon PMN activation. Therefore, increased plasma concentrations of myeloperoxidase [219-225] might reflect the physiologic answer of the cellular fibrinolysis towards a thrombus. Unfortunately, this answer often is insufficient or frustrating, and the ischemia damages the respective organ [226].

If it is difficult to selectively lyse a pathologic thrombus via individually adjusted therapeutic activation of plasmatic fibrinolysis [173] (with about 0.1-0.2 U/ml plasmatic plasmin [174]), the PMN- mediated selective fibrinolysis might be stimulated: the PMN with their CD11b/CD18 receptors adhere selectively to pathologic thrombi, the PMN get activated [227], and the PMN destroy these fibrin deposits [22,206]; CD11b/CD18 binding enhances the generation of reactive oxygen species, and CD11b/CD18 is activated via redox reactions [181-196].

Depending on the reaction conditions (presence or absence of taurine to yield taurine-chloramine) the reactive oxygen species can be physiologic ($^1O_2$*) or pathologic (·OH) [228,229]. So, pro-PMN drugs such as G-CSF [230,231] or the intravascular release of $^1O_2$*

or of photons might be an innovative therapy, e.g. to selectively activate anti-thrombotic PMN without major bleeding complications [92,147,168,208,209,213]. Hypo-activity of Nox2 may result in immunosuppression, hyper-activity of Nox2 can results in autoimmune disorders, cardiovascular diseases, or neurodegeneration [232-238]. Enzymes of the Nox family are of primary importance in regulation of blood pressure (Nox1), biosynthesis (DUOX2), and in host defence (Nox2) [238].

REFERENCES

[1]   Paape MJ, Bannerman DD, Zhao X, Lee JW. The bovine neutrophil: structure and function in blood and milk. *Vet. Res.* 2003; 34: 597-627.

[2]   Tate MD, Brooks AG, Reading PC. The role of neutrophils in the the upper and lower respiratory tract during influenza virus infection of mice. *Respir. Res.* 2008 Aug 1; 9: 57. [Epub ahead of print].

[3]   Sugui JA, Kim HS, Zarember KA, Chang YC, Gallin JI, Nierman WC, Kwon-Chung KJ. Genes differentially expressed in conidia and hyphae of Aspergillus fumigatus upon exposure to human neutrophils. *PLoS ONE* 2008; 3: e2655.

[4]   Lacy P, Eitzen G. Control of granule exocytosis in neutrophils. *Front Biosci.* 2008; 13: 5559-70.

[5]   Hogg JC. The traffic of polymorphonuclear leukocytes through pulmonary microvessels in health and disease. *Am. J. Respir.* 1994; 163: 769-778.

[6]   Dinauer MC. Disorders of neutrophil function: an overview. *Methods Mol. Biol.* 2007; 412: 489-504.

[7]   Malech HL. The role of neutrophils in the immune system: an overview. *Methods Mol. Biol* 2007; 412: 3-11.

[8]   Sawyer DW, Donowitz GR, Mandell GL.Polymorphonuclear neutrophils: an effective antimicrobial force. *Rev. Infect. Dis.* 1989; 11 Suppl 7: S1532-44.

[9]   Harlan JM. Leukocyte-endothelial interactions. *Blood* 1985; 65: 513-25.

[10]   Weiss SJ. Tissue destruction by neutrophils. *N. Engl. J. Med.* 1989; 320: 365-76.

[11]   Ellis TN, Beaman BL. Murine polymorphonuclear neutrophils produce interferon-gamma in response to pulmonary infection with Nocardia asteroids. *J. Leukocyte Biol.* 2002; 72: 373-81.

[12]   Ellis TN, Beaman BL. Interferon-gamma activation of polymorphonuclear neutrophil function. *Immunology* 2004; 112: 2-12.

[13]   Sohn EJ, Paape MJ, Connor EE, Bannerman DD, Fetterer RH, Peters RR. Bacterial lipopolysaccharide stimulates bovine neutrophil production of TNF-alpha, IL-1beta, IL-12 and IFN-gamma. *Vet. Res.* 2007; 38: 809-18.

[14]   Djeu JY. Cytokines and anti-fungal immunity. *Adv. Exp. Med. Biol.* 1992; 319: 217-23.

[15]   Sigal A, Bleijs DA, Grabovsky V, Vliet SJ, Dwir O, Figdor CG, van Kooyk Y, Alon R. The LFA-1 integrin supports rolling adhesions on ICAM-1 under physiological shear flow in a permissive cellular environment. *J. Immunol.* 2000; 165: 442-52.

[16]   Gaboury JP, Kubes P. Reductions in physiologic shear rates lead to CD11/CD18-dependent, selectin-independent leukocyte rolling in vivo. *Blood* 1994; 83: 345-50.

[17]   McEver RP. Adhesive interactions of leukocytes, platelets, and the vessel wall during hemostasis and inflammation. *Thromb. Haemost.* 2001; 86: 746-56.

[18]   Etzioni A, Tonetti M. Leukocyte adhesion deficiency II- from A to almost Z. *Immunological Reviews* 2000; 178: 138-47.

[19]   Davenpeck KL, Berens KL, Dixon RAF, Dupre B, Bochner BS. Inhibition of adhesion of human neutrophils and eosinophils to P-selectin by the sialyl Lewis^x antagonist TBC1269: preferential activity against neutrophil adhesion in vitro. *J. Allergy Clin. Immunol.* 2000; 105: 769-75.

[20]   Stie J, Jesaitis AJ. Reorganization of the human neutrophil plasma membrane is associated with functional priming: implications for neutrophil preparations. *J. Leukoc. Biol.* 2007; 81: 672-85.

[21]   Sakamoto S, Okanoue T, Itoh Y, Sakamoto K, Nishioji K, Nakagawa Y, Yoshida N, Yoshikawa T, Kashima K. Intercellular adhesion molecule-1 and CD18 are involved in neutrophil adhesion and its cytotoxicity to cultured sinusoidal endothelial cells in rats. *Hepatology* 1997; 26: 658-63.

[22]   Loike JD, Silverstein R, Wright SD, Weitz JI, Huang AJ, Silverstein SC. The role of protected extracellular compartments in interactions between leukocytes, and platelets, and fibrin/fibrinogen matrices. *Ann. N Y Acad. Sci.* 1992; 667: 163-72.

[23]   Wagner JG, Roth RA. Neutrophil migration during endotoxemia. *J. Leukoc. Biol.* 1999; 66: 10-24.

[24]   Stief TW. Coagulation Activation by Lipopolysaccharides. *Clin. Appl. Thromb. Hemost.* 2007 Dec 26.

[25]   Arditi M, Zhou J, Dorio R, Rong GW, Goyert SM, Kim KS. Endotoxin-mediated endothelial cell injury and activation: role of soluble CD14. *Infect. Immun.* 1993; 61: 3149-56.

[26]   Charavaryamath C, Janardhan KS, Caldwell S, Singh B. Pulmonary intravascular monocytes/macrophages in a rat model of sepsis. *Anat. Rec. A Discov. Mol. Cell Evol. Biol.* 2006; 288: 1259-71.

[27]   Farsky SP, Sannomiya P, Garcia-Leme J. Secreted glucocorticoids regulate leukocyte-endothelial interactions in inflammation. A direct vital microscopic study. *J. Leukoc. Biol.* 1995; 57: 379-86.

[28]   Filep JG, Delalandre A, Payette Y, Földes-Filep E. Glucocorticoid receptor regulates expression of L-selectin and CD11/CD18 on human neutrophils. *Circulation* 1997; 96: 295-301.

[29]   Hill GE, Alonso A, Thiele GM, Robbins RA. Glucocorticoids blunt neutrophil CD11b surface glycoprotein upregulation during cardiopulmonary bypass in humans. *Anesth Analg.* 1994; 79: 23-7.

[30]   Brandt MM, Lynam E, Sklar LA. Methylprednisolone inhibits three classes of neutrophil receptors in human blood in vitro. *J. Surg. Res.* 1993; 55: 632-9.

[31]   Gilbert RO, Kim CA, Yen A. Modulation, in vivo and in vitro, of surface expression of CD18 by bovine neutrophils. *Am. J. Vet Res.* 1992; 53: 1675-8.

[32]   Schramm R, Liu Q, Thorlacius H. Expression and function of MIP-2 are reduced by dexamethasone treatment in vivo. *Br. J. Pharmacol.* 2000; 131: 328-34.

[33]   Barton AF, Bayley DL, Mikami M, Llewellyn-Jones CG, Stockley RA. Phenotypic changes in neutrophils related to anti-inflammatory therapy. *Biochim. Biophys. Acta* 2000; 1500: 108-18.

[34] Burton JL, Kehrli ME Jr, Kapil S, Horst RL. Regulation of L-selectin and CD18 on bovine neutrophils by glucocorticoids: effects of cortisol and dexamethasone. *J. Leukoc. Biol.* 1995; 57: 317-25.

[35] Ballabh P, Simm M, Kumari J, Krauss AN, Jain A, Califano C, Lesser ML, Cunningham-Rundles S. Neutrophil and monocyte adhesion molecules in bronchopulmonary dysplasia, and effects of corticosteroids. *Arch. Dis. Child Fetal Neonatal. Ed* 2004; 89: F76-83.

[36] Kvietys PR, Sandig M. Neutrophil diapedesis: paracellular or transcellular? News *Physiol. Sci.* 2001; 16:15-9.

[37] Hoshi O, Ushiki T. Neutrophil extravasation in rat mesenteric venules induced by the chemotactic peptide N-formyl-methionyl-luecylphenylalanine (fMLP), with special attention to a barrier function of the vascular basal lamina for neutrophil migration. *Arch. Histol Cytol.* 2004; 67: 107-14.

[38] Lin M, Jackson P, Tester AM, Diaconu E, Overall CM, Blalock JE, Pearlman E. Matrix metalloproteinase-8 facilitates neutrophil migration through the corneal stromal matrix by collagen degradation and production of the chemotactic peptide Pro-Gly-Pro. *Am. J. Pathol.* 2008; 173: 144-53.

[39] Tschesche H, Bakowski B, Schettler A, Knäuper V, Reinke H, Krämer S. Leukodiapedesis, compartmentalisation and secretion of PMN leukocyte proteinases, and activation of PMN leukocyte procollagenase. *Adv. Exp. Med. Biol.* 1991; 297: 39-53.

[40] Faustmann PM, Dermietzel R. Extravasation of polymorphonuclear leukocytes from the cerebral microvasculature. Inflammatory response induced by alpha-bungarotoxin. *Cell Tissue Res.* 1985; 242: 399-407.

[41] Hoshi O, Ushiki T. Scanning electron microscopic studies on the route of neutrophil extravasation in the mouse after exposure to the chemotactic peptide N-formyl-methionyl-leucyl-phenylalaninphenylalanine (fMLP). *Arch. Histol. Cytol.* 1999; 62: 253-60.

[42] Michel CC, Neal CR. Openings through endothelial cells associated with increased microvascular permeability. *Microcirculation* 1999; 6: 45-54.

[43] Wolosewick JJ. Distribution of actin in migrating leukocytes in vivo. *Cell Tissue Res.* 1984; 236: 517-25.

[44] Solodkin-Szaingurten I, Levy R, Hadad N. Differential behavior of sPLA2-V and sPLA2-X in human neutrophils. *Biochim. Biophys. Acta* 2007; 1771: 155-63.

[45] Gabay JE, Scott RW, Campanelli D, Griffith J, Wilde C, Marra MN, Seeger M, Nathan CF. Antibiotic proteins of human polymorphonuclear leukocytes. *Proc. Natl. Acad. Sci. USA* 1989; 86: 5610-4.

[46] Stief TW. Polycations neutralize PAI-1. *Hemostasis Laboratory* 2008; 1: 109-18.

[47] Seguchi H, Kobayashi T. Study of NADPH oxidase-activated sites in human neutrophils. *J. Electron. Microsc.* 2002; 51: 87-91.

[48] Burgoyne RD, Morgan A. Secretory granule exocytosis. *Physiol. Rev.* 2003; 83: 581-632.

[49] Di Noto R, Lo Pardo C, Schiavone EM, Ferrara F, Manzo C, Vacca C, Del Vecchio L. All-trans retinoic acid (ATRA) and the regulation of adhesion molecules in acute myeloid leukemia. *Leuk. Lymphoma* 1996; 21: 207-9.

[50] El-Benna J, Dang PM, Gougerot-Pocidalo MA. Priming of the neutrophil NADPH oxidase activation: role of p47phox phosphorylation and NOX2 mobilization to the plasma membrane. *Semin. Immunopathol.* 2008; 30: 279-89.

[51] Stief TW. Singlet oxygen potentiates thrombolysis. *Clin. Appl. Thromb. Hemost.* 2007; 13: 259-78.

[52] Schuster DP, Brody SL, Zhou Z, Bernstein M, Arch R, Link D, Mueckler M. Regulation of lipopolysaccharide-induced increases in neutrophil glucose uptake. *Am. J. Physiol. Lung Cell Mol. Physiol.* 2007; 292: L845-51.

[53] Ha H, Lee HB. Reactive oxygen species as glucose signaling molecules in mesangial cells cultured under high glucose. *Kidney Int.* 2000; 58: 19-25.

[54] Han CH, Lee MH. Activation domain in p67phox regulates the steady state reduction of FAD in gp91phox. *J. Vet. Sci.* 2000; 1: 27-31.

[55] Shen K, Sergeant S, Hantgan RR, McPhail LC, Horita DA. Mutations in the PX-SH3A linker of p47(phox) decouple PI(3,4)P2 binding from NADPH oxidase activation. *Biochemistry.* 2008; 47: 8855-65.

[56] Babior BM. NADPH oxidase. *Curr. Opin. Immunol.* 2004; 16: 42-7.

[57] Vergnaud S, Paclet MH, El Benna J, Pocidalo MA, Morel F. Complementation of NADPH oxidase in p67-phox-deficient CGD patients: p67-phox/-40-phox interaction. *Eur. J. Biochem.* 2000; 267: 1059-67.

[58] Sheppard FR, Kelher MR, Moore EE, McLaughlin NJ, Banerjee A, Silliman CC. Structural organization of the neutrophil NADPH oxidase: phosphorylation and translocation during priming and activation. *J. Leukoc. Biol.* 2005; 78: 1025-42.

[59] Gabig TG, Crean CD, Mantel PL, Rosli R. Function of wild-type or mutant Rac2 and Rap1a GTPases in differentiated HL60 cell NADPH oxidase activation. *Blood* 1995; 85: 804-11.

[60] Yan J, Li F, Ingram DA, Quilliam LA. Rap1a is a key regulator of FGF2-induced angiogenesis and together with Rap1b controls human endothelial cell functions. *Mol. Cell Biol.* 2008; 28: 5803-10.

[61] Nathan CF. Respiratory burst in adherent human neutrophils: triggering by colony-stimulating factors CSF-GM and CSF-G. *Blood* 1989; 73: 301-6.

[62] Stief TW. Oxidized fibrin stimulates the activation of pro-urokinase and is the preferential substrate of human plasmin. *Blood Coagul Fibrinolysis* 1993; 4: 117-21.

[63] Thomas EL. Myeloperoxidase-hydrogen peroxide-chloride antimicrobial system: effect of exogenous amines on antibacterial action against *Escherichia coli. Infect. Immun.* 1979; 25: 110-6.

[64] Weiss SJ, Lampert MB, Test ST. Long-lived oxidants generated by human neutrophils: characterization and bioactivity. *Science* 1983; 222: 625-8.

[65] Weiss SJ, Klein R, Slivka A, Wei M. Chlorination of taurine by human neutrophils. Evidence for hypochlorous acid generation. *J. Clin. Invest* 1982; 70: 598-607.

[66] Schuller-Levis GB, Park E. Taurine and its chloramine: modulators of immunity. *Neurochem. Res.* 2004; 29: 117-26.

[67] Stief TW. Singlet oxygen ($^1O_2$)-oxidizable lipids in the HIV membrane, new targets for AIDS therapy? *Med. Hypotheses* 2003; 60: 575-7.

[68] Stief TW. The physiology and pharmacology of singlet oxygen. *Med. Hypotheses* 2003; 60: 567-72.

[69] Dudani AK, Martyres A, Fliss H. Short communication: rapid preparation of preventive and therapeutic whole-killed retroviral vaccines using the microbicide taurine chloramine. *AIDS Res. Hum. Retroviruses* 2008; 24: 635-42.

[70] Fürnkranz U, Nagl M, Gottardi W, Köhsler M, Aspöck H, Walochnik J. Cytotoxic activity of N-chlorotaurine on Acanthamoeba spp. *Antimicrob Agents Chemother* 2008; 52: 470-6.

[71] Gottardi W, Arnitz R, Nagl M. N-Chlorotaurine and ammonium chloride: an antiseptic preparation with strong bactericidal activity. *Int. J. Pharm.* 2007; 335: 32-40.

[72] Reeves EP, Nagl M, O'Keeffe J, Kelly J, Kavanagh K. Effect of N-chlorotaurine on Aspergillus, with particular reference to destruction of secreted gliotoxin. *J. Med. Microbiol.* 2006; 55: 913-8.

[73] Romanowski EG, Yates KA, Teuchner B, Nagl M, Irschick EU, Gordon YJ. N-chlorotaurine is an effective antiviral agent against adenovirus in vitro and in the Ad5/NZW rabbit ocular model. *Invest Ophthalmol. Vis. Sci.* 2006; 47: 2021-6.

[74] Klamt F, Shacter E. Taurine chloramine, an oxidant derived from neutrophils, induces apoptosis in human B lymphoma cells through mitochondrial damage. *J. Biol. Chem.* 2005; 280: 21346-52.

[75] Gottardi W, Nagl M. Chemical properties of N-chlorotaurine sodium, a key compound in the human defence system. *Arch. Pharm.* 2002; 335: 411-21.

[76] Gottardi W, Hagleitner M, Nagl M. The influence of plasma on the disinfecting activity of the new antimicrobial agent N-chlorotaurine-sodium in comparison with chloramine T. *J. Pharm. Pharmacol.* 2001; 53: 689-97.

[77] Nagl M, Teuchner B, Pöttinger E, Ulmer H, Gottardi W. Tolerance of N-chlorotaurine, a new antimicrobial agent, in infectious conjunctivitis - a phase II pilot study. *Ophthalmologica* 2000; 214: 111-4.

[78] Nagl M, Gottardi W. Rapid killing of Mycobacterium terrae by N-chlorotaurine in the presence of ammonium is caused by the reaction product monochloramine. *J. Pharm. Pharmacol.* 1998; 50: 1317-20.

[79] Nagl M, Pfausler B, Schmutzhard E, Fille M, Gottardi W. Tolerance and bactericidal action of N-chlorotaurine in a urinary tract infection by an omniresistant *Pseudomonas aeruginosa*. *Zentralbl Bakteriol*. 1998; 288: 217-23.

[80] Nagl M, Larcher C, Gottardi W. Activity of N-chlorotaurine against herpes simplex- and adenoviruses. *Antiviral Res.* 1998; 38: 25-30.

[81] Nagl M, Nguyen VA, Gottardi W, Ulmer H, Höpfl R. Tolerability and efficacy of N-chlorotaurine in comparison with chloramine T for the treatment of chronic leg ulcers with a purulent coating: a randomized phase II study. *Br. J. Dermatol.* 2003; 149: 590-7.

[82] Yazdanbakhsh M, Eckmann CM, Roos D. Killing of schistosomula by taurine chloramine and taurine bromamine. *Am. J. Trop Med. Hyg.* 1987; 37: 106-10.

[83] Stief TW. Modulation of granulocyte mediated thrombolysis. *Hemostasis Laboratory* 2008; 1: 77-102.

[84] Mashino T, Fridovich I. NADPH mediates the inactivation of bovine liver catalase by monochloroamine. *Arch. Biochem. Biophys.* 1988; 265: 279-85.

[85] Stief TW. Regulation of hemostasis by singlet-oxygen ($^1\Delta O_2^*$). *Curr. Vasc. Pharmacol.* 2004; 2: 357-62.

[86] Shechter Y, Burstein Y, Patchornik A. Selective oxidation of methionine residues in proteins. *Biochemistry* 1975; 14: 4497-503.

[87] Peskin AV, Winterbourn CC. Kinetics of the reactions of hypochlorous acid and amino acid chloramines with thiols, methionine, and ascorbate. *Free Radic Biol. Med.* 2001; 30: 572-9.

[88] Stief TW. The blood fibrinolysis/deep-sea analogy: a hypothesis on the cell signals singlet oxygen/photons as natural antithrombotics. *Thromb. Res.* 2000; 99: 1-20.

[89] Stief TW. The fibrinogen antigenic turbidimetric assay (FIATA). The $X^2\bar{x}$ test: the corrected chi-square comparison against the control-mean. *Clin. Appl. Thromb. Hemost* 2007; 13: 73-100.

[90] Fujimoto Y, Ikeda M, Sakuma S. Monochloramine potently inhibits arachidonic acid metabolism in rat platelets. *Biochem. Biophys. Res. Commun.* 2006; 344: 140-5.

[91] Stief TW, Jeske WP, Walenga J, Schultz C, Kretschmer V, Fareed J. Singlet oxygen inhibits agonist-induced P-selectin expression and formation of platelet aggregates. *Clin. Appl. Thromb. Hemost.* 2001; 7: 219-24.

[92] Stief TW, Fu K, Yang LH, Ramaswamy A, Fareed J. Singlet oxygen ($^1O_2$) induces selective thrombolysis in vivo by massive granulocyte infiltration into the thrombus. 43. GTH Congress, Mannheim, 24.-27.2.1999, *Ann. Hematol.* 1999; 78: A32 (FV112).

[93] Stief TW, Kurz J, Doss MO, Fareed J. Singlet oxygen inactivates fibrinogen, factor V, factor VIII, factor X, and platelet aggregation of human blood. *Thromb. Res.* 2000; 97: 473-80.

[94] Vogt W. Complement activation by myeloperoxidase products released from stimulated human polymorphonuclear leukocytes. *Immunobiology* 1996; 195: 334-46.

[95] Neher A, Nagl M, Schrott-Fischer A, Ichiki H, Gottardi W, Gunkel AR, Stephan K. N-chlorotaurine, a novel endogenous antimicrobial agent: tolerability testing in a mouse model. *Arch. Otolaryngol. Head Neck Surg.* 2001; 127: 530-3.

[96] Nagl M, Miller B, Daxecker F, Ulmer H, Gottardi W. Tolerance of N-chlorotaurine, an endogenous antimicrobial agent, in the rabbit and human eye--a phase I clinical study. *J. Ocul. Pharmacol. Ther.* 1998; 14: 283-90.

[97] Shanmugham LN, Petrarca C, Castellani ML, Symeonidou I, Frydas S, Vecchiet J, Falasca K, Tetè S, Conti P, Salini V. IL-1beta induces alkaline phosphatase in human phagocytes. *Arch. Med. Res.* 2007; 38: 39-44.

[98] Rada B, Hably C, Meczner A, Timár C, Lakatos G, Enyedi P, Ligeti E. Role of Nox2 in elimination of microorganisms. *Semin. Immunopathol.* 2008; 30: 237-53.

[99] Bengtsson SH, Gulluyan LM, Dusting GJ, Drummond GR. Novel isoforms of NADPH oxidase in vascular physiology and pathophysiology. *Clin. Exp. Pharmacol. Physiol.* 2003; 30-54.

[100] Rada B, Leto TL. Oxidative innate immune defenses by Nox/Duox family NADPH oxidases. *Contrib. Microbiol.* 2008, 15. 164-87.

[101] Schäppi MG, Jaquet V, Belli DC, Krause KH. Hyperinflammation in chronic granulomatous disease and anti-inflammatory role of the phagocyte NADPH oxidase. *Semin. Immunopathol.* 2008; 30: 255-71.

[102] Miyoshi T, Matsumoto AH, Shi W. Paradoxical increase in LDL oxidation by endothelial cells from an atherosclerosis-resistant mouse strain. *Atherosclerosis* 2007; 192: 259-65.

[103] Calingasan NY, Gibson GE. Vascular endothelium is a site of free radical production and inflammation in areas of neuronal loss in thiamine-deficient brain. *Ann. N Y Acad. Sci.* 2000; 903: 353-6.

[104] Tak PP, Zvaifler NJ, Green DR, Firestein GS. Rheumatoid arthritis and p53: how oxidative stress might alter the course of inflammatory diseases. *Immunology Today* 2000; 21: 78-82.

[105] O'Brien NC, Charlton B, Cowden WB, Willenborg DO. Nitric oxide plays a critical role in the recovery of Lewis rats from experimental autoimmune encephalomyelitis and the maintenance of resistance to reinduction. *J. Immunol.* 1999; 163: 6841-7.

[106] Matheny HE, Deem TL, Cook-Mills JM. Lymphocyte migration through monolayers of endothelial cell lines involves VCAM-1 signaling via endothelial cell NADPH Oxidase. *J. Immunol.* 2000; 164: 6550-9.

[107] Wyche KE, Wang SS, Griendling KK, Dikalov SI, Austin H, Rao S, Fink B, Harrison DG, Zafari AM. C242T CYBA polymorphism of the NADPH oxidase is associated with reduced respiratory burst in human neutrophils. *Hypertension* 2004; 43: 1246-51.

[108] Stasia MJ, Li XJ. Genetics and immunopathology of chronic granulomatous disease. *Semin. Immunopathol.* 2008; 30: 209-35.

[109] Kirk EA, Dinauer MC, Rosen H, Chait A, Heinecke JW, LeBoeuf RC. Impaired superoxide production due to a deficiency in phagocyte NADPH oxidase fails to inhibit atherosclerosis in mice. *Arterioscler. Thromb. Vasc. Biol.* 2000; 20: 1529-35.

[110] Bach TL, Barsigian C, Yaen CH, Martinez J. Endothelial cell VE-cadherin functions as a receptor for the ß15-42 sequence of fibrin. *J. Biol. Chem.* 1998; 273: 30719-28.

[111] Cockwell P, Tse WY, Savage CO. Activation of endothelial cells in thrombosis and vasculitis. *Scand. J. Rheumatol.* 1997; 26: 145-50.

[112] Ruf A, Schlenk RF, Maras A, Morgenstern E, Patscheke H. Contact-induced neutrophil activation by platelets in human cell suspensions and whole blood. *Blood* 1992; 80: 1238-46.

[113] Vercellotti GM, Severson SP, Duane P, Moldow CF. Hydrogen peroxide alters signal transduction in human endothelial cells. *J. Lab. Clin. Med.* 1991; 117: 15-24.

[114] Kirton CM, Nash GB. Activated platelets adherent to an intact endothelial cell monolayer bind flowing neutrophils and enable them to transfer to the endothelial surface. *J. Lab. Clin. Med.* 2000; 136: 303-13.

[115] Trezzini C, Schalch L, Jungi TW. The triggering of fibrinogen receptors on mononuclear phagocytes of healthy subjects and of a thrombasthenia Glanzmann patient promotes the generation of reactive oxygen species. *Blood Coagul. Fibrinolysis* 1990; 1: 363-73.

[116] Jungi TW, Spycher MO, Nydegger UE, Barandum S. Platelet-leukocyte interaction: selective binding of thrombin-stimulated platelets to human monocytes, polymorphonuclear leukocytes and related cell lines. *Blood* 1986; 67: 629-36.

[117] Sherman LA, Lee J. Specific binding of soluble fibrin to macrophages. *J. Exp. Med.* 1977; 145: 76-85.

[118] Colvin RB. Fibrinogen-fibrin interactions with fibroblasts and macrophages. *Ann. NY Acad. Sci.* 1983; 408: 621-633.

[119] Neelamegham S, Taylor AD, Shankaran H, Smith CW, Simon SI. Shear and time-dependent changes in Mac-1, LFA-1, and ICAM-3 binding regulate neutrophil homotypic adhesion. *J. Immunol.* 2000; 164: 3798-805.

[120] Rubel C, Fernandez GC, Dran G, Bompadre MB, Isturiz MA, Palermo MS. Fibrinogen promotes neutrophil activation and delays apoptosis. *J. Immunol.* 2001; 166: 2002-10.

[121] Brennan ML, Anderson MM, Shih DM, Qu XD, Wang X, Mehta AC, Lim LL, Shi W, Hazen SL, Jacob JS, Crowley JR, Heinecke JW, Lusis AJ. Increased atherosclerosis in myeloperoxidase-deficient mice. *J. Clin. Invest.* 2001; 107: 419-30.

[122] Almekhlafi MA, Hu WY, Hill MD, Auer RN. Calcification and endothelialization of thrombi in acute stroke. *Ann. Neurol.* 2008; 64: 344-8.

[123] Järemo P, Nilsson O. Interleukin-6 and neutrophils are associated with long-term survival after acute myocardial infarction. *Eur. J. Intern. Med.* 2008; 19: 330-3.

[124] Shepherd J, Hilderbrand SA, Waterman P, Heinecke JW, Weissleder R, Libby P. A fluorescent probe for the detection of myeloperoxidase activity in atherosclerosis-associated macrophages. *Chem. Biol.* 2007; 14: 1221-31.

[125] McMillen TS, Heinecke JW, LeBoeuf RC. Expression of human myeloperoxidase by macrophages promotes atherosclerosis in mice. *Circulation* 2005; 111: 2798-804.

[126] Aviram M, Rosenblat M. Paraoxonases 1, 2, and 3, oxidative stress, and macrophage foam cell formation during atherosclerosis development. *Free Radic. Biol. Med.* 2004; 37: 1304-16.

[127] Nakayama K, Nitto T, Inoue T, Node K. Expression of the cytochrome P450 epoxygenase CYP2J2 in human monocytic leukocytes. *Life Sci.* 2008; 83: 339-45.

[128] Lambert JM, Lopez EF, Lindsey ML. Macrophage roles following myocardial infarction. *Int. J. Cardiol.* 2008; 30: 147-58.

[129] van Leeuwen M, Gijbels MJ, Duijvestijn A, Smook M, van de Gaar MJ, Heeringa P, de Winther MP, Tervaert JW. Accumulation of myeloperoxidase-positive neutrophils in atherosclerotic lesions in LDLR-/- mice. *Arterioscler. Thromb. Vasc. Biol.* 2008; 28: 84-9.

[130] Cadroy Y, Dupouy D, Boneu B, Plaisancie H. Polymorphonuclear leukocytes modulate tissue factor production by mononuclear cells: role of reactive oxygen species. *J. Immunol.* 2000; 164: 3822-8.

[131] Mitarai T. Anti-neutrophil cytoplasmic autoantibodies associated renal diseases and thrombotic microangiopathy. *Intern. Med.* 1998; 37: 207-10.

[132] Tervaert JW, Stegeman CA, Kallenberg CG. Novel therapies for anti-neutrophil cytoplasmic antibody-associated vasculitis. *Curr. Opin. Nephrol. Hypertens* 2001; 10: 211-7.

[133] Liu X, Piela-Smith TH. Fibrin(ogen)-induced expression of ICAM-1 and chemokines in human synovial fibroblasts. *J. Immunol.* 2000; 165: 5255-61.

[134] Anstee QM, Goldin RD, Wright M, Martinelli A, Cox R, Thursz MR. Coagulation status modulates murine hepatic fibrogenesis: implications for the development of novel therapies. *J. Thromb. Haemost.* 2008; 6: 1336-43.

[135] Van Giezen JJJ, Chung A Hing JE, Vegter CB, Bouma BN, Jansen JWCM. Fibrinolytic activity in blood is distributed over a cellular and the plasma fraction which can be modulated separately. *Thromb. Haemost.* 1994; 72: 887-92.

[136] Moir E, Booth NA, Bennett B, Robbie LA. Polymorphonuclear leucocytes mediate endogenous thrombus lysis via a u-PA – dependent mechanism. *Br J Haematol.* 2001; 113: 72-80.

[137] Stief TW, Fröhlich S, Renz H. Determination of the global fibrinolytic state. *Blood Coagul. Fibrinolysis* 2007; 18: 479-87.

[138] Stief TW. Singlet oxygen enhances intrinsic thrombolysis: the intrinsic oxidative clot lysis assay (INOXCLA). *Clin. Appl. Thromb. Hemost.* 2007; 13: 369-83.

[139] Haj MA, Neilly IJ, Robbie LA; Adey GD, Bennett B. Influence of white blood cells on the fibrinolytic response to sepsis: studies of septic patients with or without severe leucopenia. *Br. J. Haematol.* 1995; 90: 541-7.

[140] Bach-Gansmo ET, Halvorsen S, Godal HC, Skjönsberg OH. D-Dimers are degraded by human neutrophil elastase. *Thromb. Res.* 1996; 82: 177-86.

[141] Brown KA, Brain SD, Pearson JD, Edgeworth JD, Lewis SM, Treacher DF. Neutrophils in development of multiple organ failure in sepsis. *Lancet* 2006; 368: 157-69.

[142] Stief TW, Ijagha O, Weiste B, Herzum I, Renz H, Max M. Analysis of hemostasis alterations in sepsis. *Blood Coagul. Fibrinolysis* 2007; 18: 179-86.

[143] Andrade J, Al Ali A, Saw J, Wong GC. Acute stent thrombosis in a patient with giant cell arteritis. *Can. J. Cardiol.* 2008; 24: e25-6.

[144] Menon SP, Rajkumar SV, Lacy M, Falco P, Palumbo A. Thromboembolic events with lenalidomide-based therapy for multiple myeloma. *Cancer* 2008; 112: 1522-8.

[145] Jilma B, Cvitko T, Winter-Fabry A, Petroczi K, Quehenberger P, Blann AD. High dose dexamethasone increases circulating P-selectin and von Willebrand factor levels in healthy men. *Thromb. Haemost.* 2005; 94: 797-801.

[146] Frank RD, Altenwerth B, Brandenburg VM, Nolden-Koch M, Block F. Effect of intravenous high-dose methylprednisolone on coagulation and fibrinolysis markers. *Thromb. Haemost* 2005; 94: 467-8.

[147] Stief TW. Nonradical excited oxygen species induce selective thrombolysis in vivo. *Thromb. Res.* 1991; 62: 147-63.

[148] Falanga A, Marchetti M, Evangelista V, Manarini S, Oldani E, Giovanelli S, Galbusera M, Cerletti C, Barbui T. Neutrophil Activation and hemostatic changes in healthy donors receiving granulocyte colony-stimulating factor. *Blood* 1999; 93: 2506-14.

[149] Stief T. G-CSF enhances cellular fibrinolysis. *Clin. Appl. Thromb. Hemost.* 2006; 12: 122.

[150] Kleemann R, Zadelaar S, Kooistra T. Cytokines and atherosclerosis: a comprehensive review of studies in mice. *Cardiovasc. Res.* 2008; 79: 360-76.

[151] Gyetko MR, Sud S, Kendall T, Fuller JA, Newstead MW, Standiford TJ. Urokinase receptor-deficient mice have impaired neutrophil recruitment in response to pulmonary pseudomonas aeruginose infection. *J. Immunol.* 2000; 165: 1513-9.

[152] Plow EF, Felez J, Miles LA. Cellular regulation of fibrinolysis. *Thromb. Haemost.* 1991; 66: 32-6.

[153] Vinazzer HA. Antithrombin III in shock and disseminated intravascular coagulation. *Clin. Appl. Thromb. Hemost.* 1995; 1: 62-5.

[154] Stief TW. Thrombin generation by exposure of blood to endotoxin: a simple model to study disseminated intravascular coagulation. *Clin. Appl. Thromb. Hemost.* 2006; 12: 137-61.

[155] Kelvin DJ, Michiel DF, Johnston JA, Lloyd AR, Sprenger H, Oppenheim JJ, Wang JM. Chemokines and serpentines: the molecular biology of chemokine receptors. *J. Leukoc. Biol.* 1993; 54: 604-12.

[156] Wang ZY, Chen Z. Acute promyelocytic leukemia: from highly fatal to highly curable. *Blood* 2008; 111: 2505-15.

[157] Clagett-Dame M, McNeill EM, Muley PD. Role of all-trans retinoic acid in neurite outgrowth and axonal elongation. *J. Neurobiol.* 2006; 66: 739-56.

[158] Lane MA, Bailey SJ. Role of retinoid signalling in the adult brain. *Prog. Neurobiol.* 2005; 75: 275-93.

[159] Salih HR, Kiener PA. Alterations in Fas (CD 95/Apo-1) and Fas ligand (CD178) expression in acute promyelocytic leukemia during treatment with ATRA. *Leuk. Lymphoma* 2004; 45: 55-9.

[160] Capitani S, Marchisio M, Neri LM, Brugnoli F, Gonelli A, Bertagnolo V. Phosphoinositide 3-kinase is associated to the nucleus of HL-60 cells and is involved in their ATRA-induced granulocytic differentiation. *Eur. J. Histochem.* 2000; 44: 61-5.

[161] Hofmann SL. Retinoids--"differentiation agents" for cancer treatment and prevention. *Am. J. Med. Sci.* 1992; 304: 202-13.

[162] Warso MA, Lands WE. Presence of lipid hydroperoxide in human plasma. *J. Clin. Invest.* 1985; 75: 667-71.

[163] Petrone WF, English DK, Wong K, McCord JM. Free radicals and inflammation: superoxide-dependent activation of a neutrophil chemotactic factor in plasma. *Proc. Natl. Acad. Sci. USA* 1980; 77: 1159-63.

[164] McCord JM. A superoxide-activated chemotactic factor and its role in the inflammatory process. *Agents Actions* 1980; 10: 522-7.

[165] McCord JM.Superoxide radical: controversies, contradictions, and paradoxes. *Proc. Soc. Exp. Biol. Med.* 1995; 209: 112-7.

[166] Khan AU, Kasha M. Singlet molecular oxygen in the Haber-Weiss reaction. *Proc. Natl. Acad. Sci. USA* 1994; 91: 12365-7.

[167] McCord JM, Day ED Jr. Superoxide-dependent production of hydroxyl radical catalyzed by iron-EDTA complex. *FEBS Lett.* 1978; 86: 139-42.

[168] Salin ML, McCord JM. Superoxide dismutases in polymorphonuclear leukocytes. *J. Clin. Invest* 1974; 54: 1005-9.

[169] Sluiter W, de Vree WJ, Pietersma A, Koster JF. Prevention of late lumen loss after coronary angioplasty by photodynamic therapy: role of activated neutrophils. *Mol. Cell Biochem.* 1996; 157: 233-8.

[170] Klebanoff SJ, Locksley RM, Jong EC, Rosen H. Oxidative response of phagocytes to parasite invasion. *Ciba Found Symp.* 1983; 99: 92-112.

[171] Rosen H, Klebanoff SJ. Bactericidal activity of a superoxide anion-generating system. A model for the polymorphonuclear leukocyte. *J. Exp. Med.* 1979; 149: 27-39.

[172] Rosen H, Klebanoff SJ. Formation of singlet oxygen by the myeloperoxidase-mediated antimicrobial system. *J. Biol. Chem.* 1977; 252: 4803-10.

[173] Merhi Y, King M, Guidoin R. Acute thrombogenicity of intact and injured natural blood conduits versus synthetic conduits: neutrophil, platelet, and fibrin(ogen) adsorption under various shear-rate conditions. *J. Biomed. Mater. Res.* 1997; 34: 477-85.

[174] Collen D, Lijnen HR. The fibrinolytic system in man. *Crit. Rev. Oncol. Hematol.* 1986; 4: 249-301.

[175] Stief TW, Richter A, Bünder R, Maisch B, Renz H. Monitoring of plasmin and plasminogen activator activity in blood of patients under fibrinolytic treatment by reteplase. *Clin. Appl. Thromb. Hemost.* 2006; 12: 213-8.

[176] Kwaan HC, Astrup T. Demonstration of cellular fibrinolytic activity by the histochemical fibrin slide technique. *Lab. Invest.* 1967; 17: 140-5.

[177] Langleben D, Moroz LA. Interspecies variation in the cellular phase of blood fibrinolytic activity. *Proc. Soc. Exp. Biol. Med.* 1991; 196: 270-2.

[178] Humphries JE, Vasudevan J, Gonias SL. Fibrinogenolytic and fibrinolytic activity of cell-associated plasmin. *Arterioscler. Thromb.* 1993; 13: 48-55.

[179] Calderwood JW, Williams JM, Morgan MD, Nash GB, Savage CO. ANCA induces beta2 integrin and CXC chemokine-dependent neutrophil-endothelial cell interactions that mimic those of highly cytokine-activated endothelium. *J. Leukoc. Biol.* 2005; 77: 33-43.

[180] Broudy VC, Kaushansky K, Harlan JM, Adamson JW. Interleukin 1 stimulates human endothelial cells to produce granulocyte-macrophage colony-stimulating factor and granulocyte colony-stimulating factor. J. *Immunol.* 1987; 139: 464-8.

[181] Kitagawa Y, Van Eeden SF, Redenbach DM, Daya M, Walker BA, Klut ME, Wiggs BR, Hogg JC. Effect of mechanical deformation on structure and function of polymorphonuclear leukocytes. *J. Appl. Physiol.* 1997; 82: 1397-405.

[182] Ramamoorthy C, Kovarik WD, Winn RK, Harlan JM, Sharar SR. Neutrophil adhesion molecule expression is comparable in perinatal rabbits and humans. *Anesthesiology* 1997; 86: 420-7.

[183] Schwartz BR, Harlan JM. Sulfhydryl reducing agents promote neutrophil adherence without increasing surface expression of CD11b/CD18 (Mac-1, Mo1). *Biochem. Biophys. Res. Commun.* 1989; 165: 51-7.

[184] Vedder NB, Harlan JM. Increased surface expression of CD11b/CD18 (Mac-1) is not required for stimulated neutrophil adherence to cultured endothelium. *J. Clin. Invest* 1988; 81: 676-82.

[185] Rochon YP, Kavanagh TJ, Harlan JM. Analysis of integrin (CD11b/CD18) movement during neutrophil adhesion and migration on endothelial cells. *J. Microsc.* 2000; 197: 15-24.

[186] Nagata M, Yamamoto H, Shibasaki M, Sakamoto Y, Matsuo H. Hydrogen peroxide augments eosinophil adhesion via beta2 integrin. *Immunology* 2000; 101: 412-8.

[187] Lu H, Ballantyne C, Smith CW. LFA-1 (CD11a/CD18) triggers hydrogen peroxide production by canine neutrophils. *J. Leukoc. Biol.* 2000; 68: 73-80.

[188] Lu H, Youker K, Ballantyne C, Entman M, Smith CW. Hydrogen peroxide induces LFA-1-dependent neutrophil adherence to cardiac myocytes. *Am. J. Physiol. Heart Circ. Physiol.* 2000; 278: H835-42.

[189] Blouin E, Halbwachs-Mecarelli L, Rieu P. Redox regulation of beta2-integrin CD11b/CD18 activation. *Eur. J. Immunol.* 1999; 29: 3419-31.

[190] Repo H, Rochon YP, Schwartz BR, Sharar SR, Winn RK, Harlan JM. Binding of human peripheral blood polymorphonuclear leukocytes to E-selectin (CD62E) does not promote their activation. *J. Immunol.* 1997; 159: 943-51.

[191] Fraticelli A, Serrano CV Jr, Bochner BS, Capogrossi MC, Zweier JL. Hydrogen peroxide and superoxide modulate leukocyte adhesion molecule expression and leukocyte endothelial adhesion. *Biochim. Biophys Acta* 1996; 1310: 251-9.

[192] Wikman A, Lundahl J, Fernvik E, Shanwell A. Altered expression of adhesion molecules (L-selectin and Mac-1) on granulocytes during storage. *Transfusion* 1994; 34: 167-71.

[193] Vaporciyan AA, Ward PA. Enhanced generation of $\cdot O_2^-$ by human neutrophils via a complement iC3b/Mac-1 interaction. *Biol. Signals* 1993; 2: 126-35.

[194] Shappell SB, Toman C, Anderson DC, Taylor AA, Entman ML, Smith CW. Mac-1 (CD11b/CD18) mediates adherence-dependent hydrogen peroxide production by human and canine neutrophils. *J. Immunol.* 1990; 144: 2702-11.

[195] Skoglund G, Cotgreave I, Rincon J, Patarroyo M, Ingelman-Sundberg M. H_2O_2 activates CD11b/CD18-dependent cell adhesion. *Biochem. Biophys. Res. Commun.* 1988; 157: 443-9.

[196] Berton G, Lowell CA. Integrin signalling in neutrophils and macrophages. *Cell Signal* 1999; 11: 621-35.

[197] Wang Q, Doerschuk CM. Neutrophil-induced changes in the biomechanical properties of endothelial cells: roles of ICAM-1 and reactive oxygen species. *J. Immunol.* 2000; 164: 6487-94.

[198] Darley-Usmar VM, Lelchuk R, O'Leary VJ, Knowles M, Rogers MV, Severn A. Oxidation of low-density lipoprotein and macrophage derived foam cells. *Biochem. Soc. Trans* 1990; 18: 1064-6.

[199] Aleksandrowicz P, Wolf K, Falzarano D, Feldmann H, Seebach J, Schnittler H. Viral haemorrhagic fever and vascular alterations. *Hamostaseologie* 2008; 28: 77-84.

[200] Naito M. Macrophage differentiation and function in health and disease. *Pathol. Int.* 2008; 58: 143-55.

[201] Wynn TA. Cellular and molecular mechanisms of fibrosis. *J. Pathol.* 2008; 214: 199-210.

[202] Allavena P, Sica A, Solinas G, Porta C, Mantovani A. The inflammatory microenvironment in tumor progression: the role of tumor-associated macrophages. *Crit. Rev. Oncol. Hematol.* 2008; 66: 1-9.

[203] Papatheodorou L, Weiss N. Vascular oxidant stress and inflammation in hyperhomocysteinemia. *Antioxid. Redox. Signal* 2007; 9: 1941-58.

[204] Meerarani P, Moreno PR, Cimmino G, Badimon JJ. Atherothrombosis: role of tissue factor; link between diabetes, obesity and inflammation. *Indian J. Exp. Biol.* 2007; 45: 103-10.

[205] Croce K, Libby P. Intertwining of thrombosis and inflammation in atherosclerosis. *Curr. Opin. Hematol.* 2007; 14: 55-61.

[206] Serruys PW, García-García HM, Buszman P, Erne P, Verheye S, Aschermann M, Duckers H, Bleie O, Dudek D, Bøtker HE, von Birgelen C, D'Amico D, Hutchinson T, Zambanini A, Mastik F, van Es GA, van der Steen AF, Vince DG, Ganz P, Hamm CW, Wijns W, Zalewski A; for the Integrated Biomarker and Imaging Study-2 Investigators. Effects of the Direct Lipoprotein-Associated Phospholipase A2 Inhibitor Darapladib on Human Coronary Atherosclerotic Plaque. *Circulation* 2008; 118: 1172-82.

[207] Witting PK, Zeng B, Wong M, McMahon AC, Rayner BS, Sapir AJ, Lowe HC, Freedman SB, Brieger DB. Polymorphonuclear leukocyte phagocytic function increases in plasminogen knockout mice. *Thromb. Res.* 2008; 122: 674-82.

[208] Bouly JP, Schleicher E, Dionisio-Sese M, Vandenbussche F, Van Der Straeten D, Bakrim N, Meier S, Batschauer A, Galland P, Bittl R, Ahmad M. Cryptochrome blue light photoreceptors are activated through interconversion of flavin redox states. *J. Biol. Chem.* 2007; 282: 9383-91.

[209] Hoang N, Schleicher E, Kacprzak S, Bouly JP, Picot M, Wu W, Berndt A, Wolf E, Bittl R, Ahmad M. Human and Drosophila cryptochromes are light activated by flavin photoreduction in living cells. *PLoS Biol.* 2008; 6: e160.

[210] Stief TW. Hemostasis tolerable singlet oxygen as a perspective in AIDS therapy. *Hemostasis Laboratory* 2008; 1: 21-40.

[211] Stief TW. Circulating thrombin activity in unselected routine plasmas. *Hemostasis Laboratory* 2010; 3: 7-16.

[212] Stief TW, Ulbricht K, Max M. Circulating thrombin activity in sepsis. *Hemostasis Laboratory* 2009; 2: 293-306.

[213] Stief TW, Fareed J. The antithrombotic factor singlet oxygen/light ($^1O_2$/hv). *Clin. Appl. Thromb. Hemost.* 2000; 6: 22-30.

[214] Okajima K, Yang WP, Okabe H, Inoue M, Takatsuki K. Role of leukocytes in the activation of intravascular coagulation in patients with septicaemia. *Am J Hematol.* 1991; 36: 265-71.

[215] Nosaka M, Ishida Y, Kimura A, Kondo T. Time-dependent appearance of intrathrombus neutrophils and macrophages in a stasis-induced deep vein thrombosis model and its application to thrombus age determination. *Int J Legal Med.* 2009; 123: 235-40.

[216] Tavora FR, Ripple M, Li L, Burke AP. Monocytes and neutrophils expressing myeloperoxidase occur in fibrous caps and thrombi in unstable coronary plaques. *BMC Cardiovasc Disord.* 2009; Jun 23; 9: 27

[217] Houard X, Ollivier V, Loudec L, Michel JB, Bäck M. Differential inflammatory activity across human abdominal aortic aneurysms reveals neutrophil-derived leukotriene B4 as a major chemotactic factor released from the intraluminal thrombus. *FASEB J.* 2009; 23: 1376-83.

[218] Leclerq A, Houard X, Philippe M, Ollivier V, Sebbag U, Meilhac O, Michel JB. Involvement of intraplaque hemorrhage in atherothrombosis evolution via neutrophil protease enrichment. *J Leukoc Biol.* 2007; 82: 1420-9.

[219] Mitchell AM, Nordenholz KE, Kline JA. Tandem measurement of D-dimer and myeloperoxidase or C-reactive protein to effectively screen for pulmonary embolism in the emergency department. *Acad Emerg Med.* 2008; 15: 800-5.

[220] Loria V, Dato I, Graziani F, Biasucci LM. Myeloperoxidase: a new biomarker of inflammation in ischemic heart disease and acute coronary syndromes. *Mediators Inflamm.* 2008; 2008: 135625.

[221] Aminian A, Boudjeltia KZ, Babar S, Van Antwerpen P, Lefebvre P, Crasset V, Leone A, Ducobu J, Friart A, Vanhaerverbeek M. Coronary stenting is associated with an acute increase in plasma myeloperoxidase in stable angina patients but not in patients with acute myocardial infarction. Eur J Intern Med. 2009; 20: 527-32.

[222] Chang LT, Chua S, Sheu JJ, Wu CJ, Yeh KH, Yang CH, Yip HK. Level and prognostic value of serum myeloperoxidase in patients with acute myocardial infarction undergoing primary percutaneous coronary intervention. *Circ J.* 2009; 73: 726-31.

[223] Samimi-Fard S, Dominguez-Rodriguez A, Abreu-Gonzalez P, Enjuanes-Grau C, Blanco-Palacios G, Hernandez-Baldomero IF, Bosa-Ojeda F, Marrero-Rodríguez-F. Role of myeloperoxidase as predictor of systemic inflammatory response syndrome in patients with ST-segment elevation myocardial infarction after primary percutaneous coronary intervention. *Am J Cardiol.* 2009; 104: 634-7.

[224] Tang WH, Katz R, Brennan ML, Aviles RJ, Tracy RP, Psaty BM, Hazen SL. Usefulness of myeloperoxidase levels in healthy elderly subjects to predict risk of developing heart failure. *Am J Cardiol.* 2009; 103: 1269-74.

[225] Dominguez-Rodriguez A, Abreu-Gonzalez P, Kaski JC. Diurnal variation of circulating myeloperoxidase levels in patients with ST-segment elevation myocardial infarction. *Int J Cardiol.* 2009; 144: 407-9.

[226] Goldmann BU, Rudolph V, Rudolph TK, Holle AK, Hillbebrandt M, Meinertz T, Baldus S. Neutrophil activation precedes myocardial injury in patients with acute myocardial infarction. *Free Radic Biol Med.* 2009; 47: 79-83.

[227] Boehm TK, DeNardin E. Fibrinogen binds IgG antibody and enhances IgG-mediated phagocytosis. *Hum Antibodies.* 2008; 17: 45-56.

[228] Bonomini F, Tengattini S, Fabiano A, Bianchi R, Rezzani R. Atherosclerosis and oxidative stress. *Histol Histopathol.* 2008; 23: 381-90.

[229] Stief TW. Hemostasis and infection/inflammation. *Hemostasis Laboratory* 2008; 1: 1-2.

[230] Stief T. G-CSF enhances cellular fibrinolysis. *Clin Appl Thromb Hemost.* 2006; 12: 122.

[231] Carrão AC, Chilian WM, Yun J, Kolz C, Rocic P, Lehmann K, van den Wijngaard JP, van Horssen P, Spaan JA, Ohanyan V, Pung YF, Buschmann I. Stimulation of coronary collateral growth by granulocyte stimulating factor. Role of reactive oxygen species. *Arterioscler Thromb Vasc Biol.* 2009; 29: 1817-22.

[232] Pal R, Thakur PB, Li S, Minard C, Rodney GG. Real-time imaging of NADPH oxidase activity in living cells using a novel fluorescent protein reporter. PLoS One; 2013; 8: e63989.

[233] Han W, Li H, Segal BH, Blackwell TS. Bioluminescence imaging of NADPH oxidase activity in different animal models. J Vis Exp. 2012; doi: 10.3791/3925

[234] Hernandes MS, Britto LRG. NADPH oxidase and neurodegeneration. *Current neuropharmacology.* 2012; 10: 321-7.

[235] Miyano K, Sumimoto H. Role of the small GTPase Rac in p22phox-dependent NADPH oxidases. *Biochimie.* 2007; 89: 1133-44.

[236] Manea A. NAPH oxidase-derived reactive oxygen species: involvement in vascular physiology and pathology. *Cell Tissue Res.* 2010; 342: 325-39.

[237] Pendyla S, Natarajan V. Redox regulation of Nox proteins. *Respir Physiol Neurobiol.* 2010; 174: 265-71.

[238] Bedard K, Krause KH. The NOX family of ROS-generating NADPH oxidases: physiology and pathophysiology. *Physiol. Rev.* 2007; 87: 245-313.

Chapter 16

HEMOSTASIS ACTIVATION IN INFLAMMATION: KALLIKREIN, THROMBIN, $^1\Delta O_2{}^*$

Hemostasis is the system of generation and destruction of thrombi. Hemostasis = coagulation + fibrinolysis. The fibrin – thrombi are generated first as micro-thrombi that grow to macro-thrombi. Micro-thrombi occlude the micro-vessels of our organism, resulting in limited cell necrosis of parts of one or more of our organs. Macro-thrombi can occlude macro-vessels such as the coronary, cerebral, or pulmonary arteries, resulting in myocardial infarction, apoplex, or pulmonary embolism, respectively. Thrombin is the key enzyme in human coagulation [1]. Thrombin often is generated via altered matrix – induced coagulation (AM-coagulation) [2], called also intrinsic coagulation or contact phase coagulation. In AM-coagulation an unphysiologic surface such as a xenobiotic (foreign substances or too high concentrations of own substances) or a foreign material (such as an artificial heart valve) folds factor 12 (F12) into F12a or pre-kallikrein into kallikrein (K) [3]. F12a and/or K initiate the intrinsic clotting cascade, generating the intrinsic ten-ase that consists of the enzyme F9a with its cofactor F8a bound to phospholipids and Ca^{2+}. The other coagulation pathway that generates an activator of F10 is TF-coagulation (TF= tissue factor (CD142) a 29.6 kDa protein) [4]. In TF-coagulation the extrinsic ten-ase is formed consisting of the enzyme F7a with its activator TF bound to phospholipids and Ca^{2+}. Both ten-ases generate F10a, that in complex with its cofactor F5a and bound to phospholipids and Ca^{2+} in the common pathway of human blood coagulation activates prothrombin (F2) to thrombin (F2a). Thrombin transforms soluble fibrinogen into an insoluble fibrin network that is the basal supporter protein of micro-thrombi. About 10-100 mg/l (30-300% of normal) soluble fibrin can circulate systemically and could be used as biomarker for coagulation activation [5]. However, the main hemostasis enzyme F2a itself entrapped in α2-macrogobulin (α2M) is the preferred biomarker for systemic coagulation activation [6,7].

The systemic F2a•α2M concentration

- does not "jump" (i.e. no sudden unexplainable dramatic increase or decrease),
- is independent on fibrinogen substrate concentrations,
- does not precipitate anywhere in the body, the sampling tube, or in the reaction well,
- has a clinically interesting half live of about 30 min (not too long, not too short),

- is eliminated from the circulation by robust reticuloendothelial macrophages or dendritic cells (no interaction with sensible eventually diseased hepatocytes),
- correlates well with the severity of systemic coagulation activation [8].

The thrombin activity is down-regulated by antithrombin-3 (AT-3)/polysulphated glycosaminoglycans (SGAG) [9,10]. The AT-3/SGAG complex is our best own "drug" against severe dysregulations of normal $\rightarrow$ pathologic systemic coagulation (NIC$\rightarrow$PIC) [11,12]. Thrombin can also be entrapped in nascent fibrin, named antithrombin-1 (AT-1) [13], an action of minor physiologic importance *in vivo* but of major importance in assays *in vitro* [14]. The AT-1 action has to be reduced to almost zero, otherwise the plasmatic thrombin activities are blunted. The AT-1 action can only be eliminated within the initial minutes of clotting, where fibrin still is not crosslinked by F13a [15]. The AT-1 generated within the first clotting minutes is non-crosslinked fibrin that can be de-polymerized by supra-1molar concentrations of arginine [16]. This is the assay principle of the RECA (recalcified coagulation activity assay) [17-19] and of the thereof derived thrombin generation assays INCA (intrinsic coagulation activity assay for AM-coagulation) [20-24] and EXCA (extrinsic coagulation activity assay for TF-coagulation) [18,25,26].

Once a micro-clot has formed, the body answers with activation of fibrinolysis, consisting of plasmatic and cellular fibrinolysis [27]. The importance of plasmatic fibrinolysis with its main enzyme plasmin is not greater than 50%. At least 50% of importance have the neutrophils (PMN) the main cells of cellular fibrinolysis [28,29]. The PMN are activated by activated clotting factors such as systemic kallikrein (K) [30,31] or fibrin [32,33] that is formed by intrinsically (via F12a/K [34-40]) or extrinsically (via F7a) generated F2a. Their main weapons are reactive oxygen species (ROS) generated by triggered membrane assembly of intracellular NADPH-oxidase. A typical trigger is zymosan A (ZyA) a cell wall component of fungi [41]. ZyA binds to the PMN receptor CD11b/18 that discrimates "non-self" from "self" [4]. PMN are primed for activation by ROS or by light quants. The most important reacting oxygen derivatives are mother ROS hydrogen peroxide (H_2O_2) and her two daughter ROS hydroxyl-radical ($\cdot OH$) and non-radicalic singlet oxygen ($^1\Delta O_2{}^*$) [42,43]. Especially $^1\Delta O_2{}^*$ is of major interest for the human organism because it selectively destroys "non-self" in physiological inflammation [44].

Coagulation activation in acute PIC is caused by cell fragments (AM-coagulation, eventually TF-coagulation, if TF rich cells such as brain or placenta cells are disrupted). Acute PIC is not caused by pathologic inflammation, acute PIC may cause pathologic inflammation [18,45]. Chronic PIC might be caused by pathologic inflammation and viceversa.

Neutrophil granulocytes (PMN) are our primary defense cells against fungi, bacteria, parasites, or micro-thrombi. Their main weapons and signals are the reactive oxygen species (ROS). H_2O_2, $\cdot OH$, $^1\Delta O_2{}^*$ generation is favoured in certain micro-ambiences (PMN myeloperoxidase, taurine, alkenes) [46]. $^1\Delta O_2{}^*$ reacts with S-H, S-S, or C=C and releases a photon (hv) in the about 300-400 nm range after reaction to excited carbonyl. The activation of the assembly of their NADPH-oxidase, one main trigger (zymosan A) and one main primer ($^1\Delta O_2{}^*$/hv) are of great physiological and pathophysiological importance in inflammation and in hemostasis. There is photonic communication: the neutrophils generate different types of photons and they can "see" them of others via short wave length (blue) opsin (OPN1SW) and ultra-short wave length (UVA) opsin (OPN5). The about 300-400 nm photons are the main signals and the photons of lowest wave length seem to especially alert them in life-threatening emergency such as severe ischemia due to micro-thrombi.

The regulation of the neutrophil's ROS generation by metabolites or drugs that generate the PMN activator kallikrein via altered matrix (AM) - coagulation activation, by singlet oxygen (the excited "pro-drug" of photons), or by photons is very important. $^1\Delta O_2^*$ releases a photon by

- dimol emission (634 nm red fluorescence, by collision of two $^1\Delta O_2^*$ molecules),
- monomol emission (1270 nm infra-red phosphorescence, spontaneous),
- excited carbonyl (R-C=O*) emission (in the about 300-400 nm spectrum, depending on R, after the ene (C=C) oxidation); the emission can be shifted red-ward by fluor or phosphor (Figures 1,2).

Figure 1. Arancina, here returning from a free meadow excursion, randomly modeling as singlet oxygen ($^1\Delta O_2^*$) with a perspectivic green photon.
Reactive and signaling singlet oxygen ($^1\Delta O_2^*$) releases a photon by ●Excited carbonyl (R-C=O*) emission (about 300-400 nm), ●Monomol emission (1270 nm), ●Dimol emission (634 nm).

The best way to analyze the global kallikrein generating capacity of xenobiotics (drugs or physiologic substances at unphysiologic concentrations) in AM-coagulation is to perform a RECA (recalcified coagulation activity assay) or a CPA (contact phase assay) [39,60-62]. Therefore, the global plasma kallikrein generation tests RECA or CPA are combined with the blood ROS generation assay (BRGA) [41,63] to measure blood modulation in hemostasis/inflammation [44].

The regulation of the generation and the perception of the light signals of circulating blood neutrophils is of great present and future importance in human medicine.

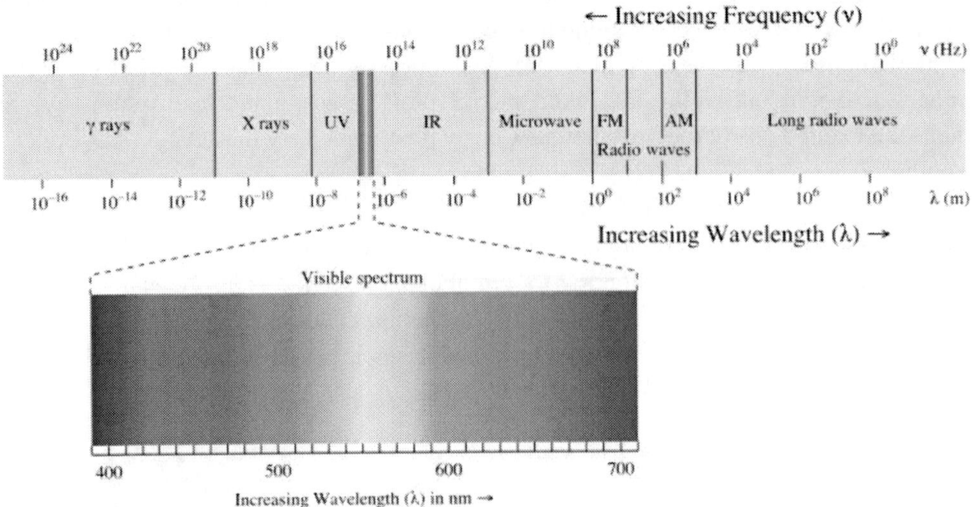

Figure 2. Electromagnetic wavelengths and the narrow visible spectrum [4].
The retina of the human eye sees only electromagnetic wavelengths in the small range of about 400-700 nm. Wavelengths of 315-380 nm are ultraviolet A (UVA) [4,47]. Typical wavelengths are for UVA 340 nm, violet 400 nm, blue 450 nm, green 530 nm, yellow 580 nm, orange 620 nm, red 700 nm [48]. UVA and violet light activates neutrophils to express their attack system consisting of ROS, proteases, phospholipase A_2 polycations, Ca^{2+} chelators [29,49-52] that together eliminate fungi, bacteria, parasites, or micro-thrombi [33,53-59].

REFERENCES

[1]   Stief TW. Thrombin – applied clinical biochemistry of the main factor of coagulation (Preface). In: *Thrombin: function and pathophysiology.* Stief T, ed.; Nova science publishers; New York; 2012; pp. vii-xx. https://www.novapublishers.com /catalog/ product_info.php?products_id=33386

[2]   Stief TW. Coumarins trigger altered matrix (AM) coagulation activation. *Hemostasis Laboratory* 2013; 6: 121-128.

[3]   Stief TW. The influence of the surface on thrombin generation. *International Journal of Laboratory Hematology* 2008; 30: 269-277.

[4]   www.wikipedia.org

[5]   Stief TW. Functional determination of soluble fibrin polymers (SFP) in plasma. *Thrombosis/Haemostasis* 2000; 84: 1120-1.

[6]   Stief TW. Specific determination of plasmatic thrombin activity. *Clin Appl Thrombosis/Haemostasis* 2006; 12: 324-9.

[7]   Stief TW. Circulating thrombin activity in unselected routine plasmas. *Hemostasis Laboratory* 2010; 3: 7-16.

[8]   Stief TW, Ulbricht K, Max M. Circulating thrombin activity in sepsis. *Hemostasis Laboratory* 2009; 2: 293-306.

[9]   Stief TW. Antithrombin III determination in nearly undiluted plasma. *Laboratory Medicine* 2008; 39: 46-8.

[10] Stief TW. Low-Molecular-Weight-Heparin (LMWH) corrects insufficient plasmatic antithrombin activity. *Hemostasis Laboratory* 2011; 4: 455-66.

[11] Stief TW. The laboratory diagnosis of the pre-phase of pathologic disseminated intravascular coagulation. *Hemostasis Laboratory* 2008; 1: 2-20.

[12] Bick RL. *Disseminated intravascular coagulation and related syndromes.* CRC Press Boca Raton, 1983.

[13] Mosesson MW. Update on antithrombin I (fibrin). *Thromb Haemost.* 2007, 98, 105-8.

[14] Stief TW. Pathological thrombin generation by the synthetic inhibitor argatroban. *Hemostasis Laboratory* 2009; 2: 83-104.

[15] Stief T, Albert A. Water quality routinely tested by RECA. *Hemostasis Laboratory* 2013; 6: 361-5.

[16] Stief TW. Inhibition of thrombin in plasma by heparin or arginine. *Clin Appl Thrombosis/Hemostasis* 2007; 13: 146-53.

[17] Stief TW. The recalcified coagulation activity. *Clin Appl Thrombosis/Hemostasis* 2008; 14: 447-53.

[18] Stief TW. Thrombin generation by exposure of blood to endotoxin – a simple model to study DIC. *Clin Appl Thrombosis/Hemostasis* 2006; 12: 137-61.

[19] Stief TW. Inhibition of thrombin generation in recalcified plasma. *Blood Coagulation and Fibrinolysis* 2007; 18: 751-60.

[20] Stief TW. Inhibition of intrinsic thrombin generation. *Drug Target Insights* 2006; 2: 6-11.

[21] Stief TW, Otto S, Renz H. The intrinsic coagulation activity assay. *Blood Coagul Fibrinolysis.* 2006; 17: 369-78.

[22] Stief TW, Otto S, Renz H. Influence of coagulation factors on intrinsic thrombin generation. *Blood Coagul Fibrinol.* 2007; 18: 67-71.

[23] Stief TW. Ellagic acid as a stable INCA-trigger. *Hemostasis Laboratory* 2009; 2: 23-32.

[24] Stief T. An "APTT" for LMWH. *Hemostasis Laboratory* 2013; 6: 367-72.

[25] Stief TW, Wieczerzak A, Renz H. Influence of coagulation factors on extrinsic thrombin generation. *Blood Coagul Fibrinolysis.* 2007; 18: 105-12.

[26] Stief TW. 20% EXCA (0.1 IU/ml thrombin) – the ideal anticoagulant target value. *Hemostasis Laboratory* 2011; 4: 275-80.

[27] Stief TW, Fröhlich S, Renz H. Determination of the global fibrinolytic state. *Blood Coagul Fibrinolysis* 2007; 18: 479-87.

[28] Stief TW. Modulation of Granulocyte Mediated Thrombolysis. *Hemostasis Laboratory* 2008; 1: 77-102.

[29] Stief TW. Neutrophil granulocytes in hemostasis. *Hemostasis Laboratory* 2008; 1: 269-89.

[30] Wachtvogel YT, Kettner C, Hack CE, Nuijens JH, Reilly TM, Knabb RM, Kucich U, Niewiarowski S, Edmunds LH Jr, Colman RW. Thrombin and human plasma kallikrein inhibition during simulated extracorporal circulation block platelet and neutrophil activation. *Thromb Haemost.* 1998; 80: 686-91.

[31] Loureiro-Silva MR, Kouyoumdjian M, Borges DR. Plasma kallikrein and thrombin are cleared through unrelated hepatic pathways. *Blood Coagul Fibrinolysis* 1993; 4: 551-4.

[32] Stief TW. Renal micro-thrombi by trimethoprim/sulfamethoxazole – a hypothesis. *Hemostasis Laboratory* 2013; 6: 107-14.

[33] Stief T. Micro-thrombi stimulate blood ROS generation. *Hemostasis Laboratory* 2013; 6: 315-25.

[34] Kaplan AP, Ghebrehiwet B. The plasma bradykinin-forming pathways and its interrelationships with complement. *Mol Immunol* 2010; 47: 2161-9.

[35] Stief TW, Klingmüller V. Diagnostic ultrasound activates pure prekallikrein. *Blood Coagulation and Fibrinolysis* 2012; 23: 781-3.

[36] Stief TW. Kallikrein activates prothrombin. *Clin Appl Thrombosis/Hemostasis* 2008: 14: 97-8.

[37] Stief TW. Kallikrein activates factor 10. *Hemostasis Laboratory* 2012; 5: 211-8.

[38] Stief TW. Kallikrein triggers thrombin generation. *Hemostasis Laboratory* 2009; 2: 45-56.

[39] Stief TW. The contact phase activity assay. *Hemostasis Laboratory* 2009; 1: 1-22.

[40] Stief TW. Zn^{2+}, hexane, or glucose activate factor 12 and/or prekallikrein in two purified systems. *Hemostasis Laboratory* 2011; 4: 409-26.

[41] Stief T. The routine blood ROS generation assay (BRGA) triggered by typical septic concentrations of zymosan A. *Hemostasis Laboratory* 2013; 6: 89-98.

[42] Stief TW. The blood fibrinolysis / deep-sea analogy: a hypothesis on the cell signals singlet oxygen/photons as natural antithrombotics. *Thromb Res.* 2000; 99: 1-20.

[43] Stief TW. The physiology and pharmacology of singlet oxygen. *Med Hypoth.* 2003; 60: 567-72.

[44] Stief TW. Hemostasis and Infection/Inflammation. *Hemostasis Laboratory* 2008; 1: 1-2.

[45] Stief TW, Max M. Active Endotoxin in Sepsis. *Hemostasis Laboratory* 2008; 1: 53-60.

[46] Weiss SJ, Lampert MB, Test ST. Long-lived oxidants generated by human neutrophils: characterization and bioactivity. *Science.* 1983; 222: 625-8.

[47] Dai T, Vrahas MS, Murray CK, Hamblin MR. Ultraviolet C irradiation: an alternative antimicrobial approach to localized infections? *Exp Rev Anti Infect Ther.* 2012; 10: 185-95. doi: 10.1586/eri.11.166.

[48] Stief T. Neutrophils' photons and opsins. In: *Photonic Hemostasis. Physiology of Light Signals in the Neutrophil.* Stief T, ed. Nova Science Publishers, New York, 2013.

[49] Henson PM, Johnston RB Jr. Tissue injury in inflammation. Oxidants, proteinases, and cationic proteins. *J Clin Invest.* 1987; 79: 669-74.

[50] Elferink JG, de Koster BM, Boonen GJ. Cytochalasin B-induced superoxide production in polycation-treated neutrophils. *Inflammation.* 1991; 15: 413-25.

[51] Ginsburg I, Misgav R, Pinson A, Varani J, Ward PA, Kohen R. Synergism among oxidants, proteinases, phospholipases, microbial hemolysins, cationic proteins, and cytokines. *Inflammation.* 1992; 16: 519-38.

[52] Grimbaldeston MA, Geczy CL, Tedla N, Finlay-Jones JJ, Hart PH. S100A8 induction in keratinocytes by ultraviolet A irradiation is dependent on reactive oxygen intermediates. *J Invest Dermatol.* 2003; 121: 1168-74.

[53] Tanaka M, Mroz P, Dai T, Huang L, Morimoto Y, Kinoshita M, Yoshihara Y, Nemoto K, Shinomya N, Seki S, et al. Photodynamic therapy can induce a protective innate immune response against murine bacterial arthritis via neutrophil accumulation. *PLoS One.* 2012; 7: e39823.

[54] Stief TW, Fu K, Yang LII, Ramaswamy A, Fareed J. Singlet oxygen ($^1O_2$) induces selective thrombolysis in vivo by massive granulocyte infiltration into the thrombus. 43. GTH Congress, Mannheim, 24.-27.2.1999, *Ann. Hematol.* 1999; 78: A32 (FV112).

[55] Romani L, Puccetti P. Controlling pathogenic inflammation to fungi. *Expert Rev Anti Infect Ther.* 2007 5: 1007-17.

[56] Donnelly RF, McCarron PA, Tunney MM, David Woolfson A. Potential of photodynamic therapy in treatment of fungal infections of the mouth. Design and characterisation of a mucoadhesive patch containing toluidine blue O. *J Photochem Photobiol B.* 2007; 86: 59-69.

[57] Dai T, Gupta A, Huang YY, Sherwood ME, Murray CK, Vrahas MS, Kielian T, Hamblin MR. Blue Light Eliminates community-acquired methicillin-resistant *staphylococcus aureus* in infected mouse skin abrasions. *Photomed Laser Surg.* 2013; 57: 1238-45.

[58] Dai T, Gupta A, Huang YY, Yin R, Murray CK, Vrahas MS, Sherwood ME, Tegos GP, Hamblin MR. Blue light rescues mice from potentially fatal Pseudomonas aeruginosa burn infection: efficacy, safety, and mechanism of action. *Antimicrob Agents Chemother.* 2013; 57: 1238-45.

[59] Huang YY, Tanaka M, Vecchio D, Garcia-Diaz M, Chang J, Morimoto Y, Hamblin MR. Photodynamic therapy induces an immune response against a bacterial pathogen. *Expert Rev Clin Immunol.* 2012; 8: 479-94.

[60] Stief TW, Mohrez M. Glucose activates human intrinsic coagulation *in vivo*. *Hemostasis Laboratory* 2012; 5: 83-9.

[61] Stief TW. Xenobiotic - induced pancreas carcinoma following mesenteric vein thrombosis: a hypothesis. *Hemostasis Laboratory* 2010; 3: 253-8.

[62] Stief TW. Thrombin triggers its generation in individual platelet poor plasma. *Hemostasis Laboratory* 2009; 2: 363-378.

[63] Stief T. Glucose initially inhibits and later stimulates blood ROS generation. *Journal of Diabetes Mellitus* 2013, 3: 15-21.

INDEX

E

F

G